Springer Theses

Recognizing Outstanding Ph.D. Research

For further volumes:
http://www.springer.com/series/8790

Aims and Scope

The series "Springer Theses" brings together a selection of the very best Ph.D. theses from around the world and across the physical sciences. Nominated and endorsed by two recognized specialists, each published volume has been selected for its scientific excellence and the high impact of its contents for the pertinent field of research. For greater accessibility to non-specialists, the published versions include an extended introduction, as well as a foreword by the student's supervisor explaining the special relevance of the work for the field. As a whole, the series will provide a valuable resource both for newcomers to the research fields described, and for other scientists seeking detailed background information on special questions. Finally, it provides an accredited documentation of the valuable contributions made by today's younger generation of scientists.

Theses are accepted into the series by invited nomination only and must fulfill all of the following criteria

- They must be written in good English.
- The topic of should fall within the confines of Chemistry, Physics and related interdisciplinary fields such as Materials, Nanoscience, Chemical Engineering, Complex Systems and Biophysics.
- The work reported in the thesis must represent a significant scientific advance.
- If the thesis includes previously published material, permission to reproduce this must be gained from the respective copyright holder.
- They must have been examined and passed during the 12 months prior to nomination.
- Each thesis should include a foreword by the supervisor outlining the significance of its content.
- The theses should have a clearly defined structure including an introduction accessible to scientists not expert in that particular field.

Alexander Grohsjean

Measurement of the Top Quark Mass in the Dilepton Final State Using the Matrix Element Method

Doctoral Thesis accepted by
University of Munich, Germany

 Springer

Author
Dr. Alexander Grohsjean
CEA Saclay Irfu/SPP 141
91191 Gif-sur-Yvette
France
e-mail: agrohsje@fnal.gov

Supervisor
Prof. Otmar Biebel
Faculty of Physics
University of Munich
Am Coulombwall 1
85748 Garching
Germany
e-mail: Otmar.Biebel@physik.uni-muenchen.de

ISSN 2190-5053

e-ISSN 2190-5061

ISBN 978-3-642-14069-3

e-ISBN 978-3-642-14070-9

DOI 10.1007/978-3-642-14070-9

Springer Heidelberg Dordrecht London New York

Library of Congress Control Number: 2010932328

Cover design: eStudio Calamar, Berlin/Figueres

Printed on acid-free paper

Springer is part of Springer Science+Business Media (www.springer.com)

Supervisor's Foreword

The main pacemakers of scientific research are curiosity, ingenuity, and a pinch of persistence. Equipped with these characteristics a young researcher will be successful in pushing scientific discoveries. And there is still a lot to discover and to understand.

In the course of understanding the origin and structure of matter it is now known that all matter is made up of six types of quarks. Each of these carry a different mass. But neither are the particular mass values understood nor is it known why elementary particles carry mass at all. One could perhaps accept some small generic mass value for every quark, but nature has decided differently. Two quarks are extremely light, three more have a somewhat typical mass value, but one quark is extremely massive. It is the top quark, the heaviest quark and even the heaviest elementary particle that we know, carrying a mass as large as the mass of three iron nuclei.

Even though there exists no explanation of why different particle types carry certain masses, the internal consistency of the currently best theory—the standard model of particle physics—yields a relation between the masses of the top quark, the so-called W boson, and the yet unobserved Higgs particle. Therefore, when one assumes validity of the model, it is even possible to take precise measurements of the top quark mass to predict the mass of the Higgs (and potentially other yet unobserved) particles.

Precise measurements of the top quark and W boson masses, paired with a discovery of the Higgs particle, could therefore lead to either a triumph of the standard model (if the prediction of the model turns out to be correct) or a hint towards which alternative theory could be the ultimately correct description of nature (if the standard model prediction turns out to be wrong). In this context the precise measurement of the top quark mass by Dr. Alexander Grohsjean is of utmost importance.

The precision of every measurement relies on the precision of the scale. A bad scale cannot yield a precise measurement. So, most effort for a mass measurement goes into setting-up a very precise scale. This is Dr. Alexander Grohsjean's merit. He implemented a very sensitive method to measure the mass of top quarks from a particular final state which shows two leptons, two neutrinos, and two jets of hadrons stemming from b quarks. His method implements the results of the theoretical calculation for the reaction into the measurement of the top quark mass. Alexander Grohsjean does not need to limit his measurement to particularly well-measured events but can include every observed final state into his measurement of the mass, e.g. also events with more than two jets. So he achieves the optimal statistical precision for the measured top quark mass. Additionally his approach allows for a decoupling of systematic uncertainties which are related to the energy measurement of the two jets of hadrons.

And there is still much more potential to Alexander Grohsjean's approach than simply measuring the top quark mass. It offers a framework that can be used to study also specific properties of top quark decays, like the helicity of the Wboson or the correlation of spins. This allows to get insight into the dynamics of the production and decay of top-antitop quark pairs. Dr. Alexander Grohsjean has done more than one step towards detailed investigations of the properties of top quarks. It is a cutting-edge result which leads beyond a mere mass determination of the top quark. He pioneered by his thesis the direction of future precision analysis at particle physics experiments in general. His measurement principle can for example be applied for precision measurements at the new Large Hadron Collider that just started its high energy proton collision programme.

Munich, May 2010   Otmar Biebel

Contents

1 Introduction 1
   References 3

2 Experimental Environment 5
   2.1 The Tevatron Accelerator Complex 5
   2.2 The DØ Experiment 7
        2.2.1 The Coordinate System 8
        2.2.2 The Tracking System 8
        2.2.3 The Calorimeter 10
        2.2.4 The Muon Spectrometer 11
        2.2.5 The Luminosity Monitors 13
        2.2.6 The Trigger System and Data Acquisition 14
   References 16

3 Event Reconstruction and Simulation 17
   3.1 Tracks 17
   3.2 Vertices 18
        3.2.1 Primary Vertex 18
        3.2.2 Secondary Vertex 18
   3.3 Electrons 18
   3.4 Muons 19
   3.5 Calibration of Charged Leptons 20
   3.6 Jets 21
   3.7 Jet Energy Scale 22
        3.7.1 Overall Jet Energy Scale 22
        3.7.2 Bottom Quark Jet Energy Scale 27
        3.7.3 Jet Energy Resolution 29
   3.8 Missing Transverse Energy $\not{E}_T$ 29
   3.9 Data Quality 29
   3.10 Monte Carlo Simulation 30
   References 30

4   The Top Quark and the Concept of Mass 33
   4.1   The Top Quark within the Standard Model. 33
   4.2   Top Quark Production 35
   4.3   Top Quark Decay 37
      4.3.1   Dilepton Channel 38
      4.3.2   Lepton + Jets Channel. 39
      4.3.3   Alljets Channel. 39
   4.4   The Concept of Mass. 39
   4.5   Relevance of the Top Quark Mass. 40
   4.6   Measurement Methods of the Top Quark Mass 41
      4.6.1   Indirect Constraints. 42
      4.6.2   Reconstruction of Decay Products. 42
      4.6.3   Measurements of the $t\bar{t}$ Cross Section 42
   References 43

5   The Matrix Element Method 45
   5.1   Event Likelihood. 45
   5.2   Process Likelihood. 46
   5.3   Description of the Detector Response. 47
      5.3.1   Parameterization of the Jet Energy Resolution 50
      5.3.2   Parameterization of the Muon Momentum Resolution 58
      5.3.3   Parameterization of the τ Lepton Decay 60
   5.4   Parameterization of the Top Pair Transverse Momentum 62
   5.5   Calculation of the Signal Likelihood 64
   5.6   Normalization of the Signal Likelihood 67
   5.7   Calculation of the Background Likelihood 68
   5.8   Normalization and Performance of the Background Likelihood. . . 70
   5.9   Likelihood Evaluation 70
   5.10 Ensemble Testing 72
   References 74

6   Measurement of the Top Quark Mass. 77
   6.1   Data Samples and Event Selection. 77
   6.2   Parton-Level Studies 81
      6.2.1   Monte Carlo Samples 86
      6.2.2   Normalization. 87
      6.2.3   Signal-Only Studies. 88
      6.2.4   Studies Including $(Z \rightarrow \tau\tau)$ jj and $(Z \rightarrow \tau\tau)$ bb Events . . . 90
      6.2.5   Studies Including $(Z \rightarrow \tau\tau)$ jj and $WWjj$ Events 93
      6.2.6   Measurement of the Signal Fraction 97
   6.3   Calibration of the Method. 97
      6.3.1   Monte Carlo Samples 99
      6.3.2   Normalization. 100

6.3.3   Validation of the Integration over the Top Pair
         Transverse Momentum   101
6.3.4   Calibration for Run IIa and Run IIb   102
6.4   Measurement.   105
6.5   Systematic Uncertainties.   107
6.5.1   Detector Modeling   108
6.5.2   Physics Modeling   110
6.5.3   Uncertainties from the Measurement Method   112
6.5.4   Summary of Systematic Uncertainties   113
6.6   Combination of the Run IIa and Run IIb Mass Measurements . .   113
6.6.1   The Best Linear Unbiased Method   113
6.6.2   Combination.   115
References   117

7   Improved Mass Measurement.   119
7.1   Motivation   119
7.2   Modifications   120
7.2.1   Event Likelihoods   121
7.2.2   Normalization of the Jet Transfer Functions.   121
7.2.3   Likelihood Evaluation   122
7.3   Parton-Level Studies   123
7.3.1   Monte Carlo Samples   123
7.3.2   Normalization.   123
7.3.3   Signal-Only Studies.   123
7.3.4   Studies Including $(Z \rightarrow \tau\tau)$ jj and $(Z \rightarrow \tau\tau)$ bb Events . .   128
7.3.5   Studies Including $(Z \rightarrow \tau\tau)$ jj and $WWjj$ Events   135
References   135

8   Conclusion.   137
8.1   Summary and Interpretation   137
8.2   Outlook   140
References   141

Appendices   143
A: Solving for the Event Kinematics   143
B: The Jacobian Determinant for the Signal Integration.   145
References   148

Chapter 1
Introduction

The discovery of the top quark in 1995 by the CDF [1] and DØ experiments [2] at the Fermilab Tevatron collider in the mass range of 173^{+18}_{-20} GeV[1] predicted from the fits to the electroweak data of the LEP and SLD experiments [3] was a great success for the Standard Model of particle physics. Of all known fundamental particles, the top quark has the largest mass. This makes it unique from a theoretical point of view. Due to its high mass, the Yukawa coupling of the top quark is close to unity, suggesting that it may play a special role in electroweak symmetry breaking [4]. Moreover, precise measurements of the top quark mass constrain the mass of the yet-unobserved Higgs boson through radiative corrections to the W boson mass and allow for a restriction of possible extensions to the Standard Model [5]. Experimentally, the top quark is one-of-a-kind due to its short lifetime of about 5×10^{-25} s [6]. The top quark is the only quark that cannot form hadrons and it therefore allows direct measurements of its properties. Various techniques are used to measure the top quark mass in different final states [7]. Their average value of

$$172.4 \pm 0.7\,(\text{stat.}) \pm 1.0\,(\text{syst.})\ \text{GeV}$$

[8] is in striking agreement with the prediction of 179^{+12}_{-9} GeV from indirect constraints within the Standard Model [9].

Up to the start of the Large Hadron Collider (LHC) in 2009, the Tevatron, the proton-antiproton collider at the Fermi National Accelerator Laboratory (Fermilab), is the only place on Earth where top quarks can be produced directly. At a center-of-mass energy of 1.96 TeV, top quarks are dominantly produced in top–antitop pairs via quark–antiquark annihilation. Both top and antitop are predicted to decay almost exclusively to a W boson and a b quark. According to the number

---

[1] Throughout this thesis, the convention $\hbar = 1, c = 1$ is followed.

A. Grohsjean, *Measurement of the Top Quark Mass in the Dilepton Final State Using the Matrix Element Method*, Springer Theses,
DOI: 10.1007/978-3-642-14070-9_1, © Springer-Verlag Berlin Heidelberg 2010

of leptonic *W* decays, top quark events are classified into dilepton, lepton + jets, and alljets.

During the so-called Run I period of the Tevatron collider, the Matrix Element method has been developed at the DØ experiment to extract the mass in the lepton + jets final state from a small data sample [10]. Still today, the Matrix Element method yields the single most precise measurement of the top quark mass [11]. In April 2006, CDF published the first measurement with this method in the dilepton channel using 340 pb^{-1} of data [12].

The Matrix Element method is based on the likelihood to observe a given event under the assumption of a certain value for the quantity to be measured, e.g. the mass of the top quark. Compared to conventional methods, the statistical uncertainty assigned to the measurement is significantly smaller. The excellent performance of the Matrix Element method can be mainly explained by two different aspects. First of all, each event is corrected back to the partonic level, i.e. for jets, both the energy resolution and the transition from the quark to the jet level are taken into account. Second, the contribution of each event to the final measurement is weighted according to the likelihood span of this event, i.e. poorly measured events contribute less than well measured events. On the other hand, this complex approach requires a lot of computing time and it is clear that it can only be used where the dataset is relatively small.

Although the world average of the top quark mass is dominated by the lepton + jets channel so far, the dilepton channel is of interest as a cross check and to search for possible differences. In addition, as systematic uncertainties become more and more important with increasing statistics, the dilepton channel has the potential to contribute significantly to the world average. It is therefore desirable to use the Matrix Element method in both the lepton + jets and the dilepton final states.

In the present thesis, the first measurement of the top quark mass in the dilepton channel at the DØ experiment is discussed. After significant improvements to the method [13, 14], a preliminary result based on 2.8 fb^{-1} of electron + muon data collected at the DØ experiment until May 2008 has been published [15].

In contrast to the lepton + jets channel with four jets, one charged lepton and one neutrino, the event topology in the dilepton channel is characterized by two jets, two charged leptons and two undetected neutrinos. Thus, the kinematic reconstruction of the final-state particles poses a special challenge in this channel. Given the small dilepton event sample, it is not desirable to restrict the analysis to events with exactly two jets, and the modeling of additional jets in the Matrix Element method has been developed. In addition, the dominant background of *Z* bosons produced in association with jets includes the decay of two τ leptons which required the design of a completely new process likelihood.

To validate the Matrix Element method Monte Carlo simulated events at the generator level are used. For the measurement, calibration curves are derived from events that are run through the full DØ detector simulation. The analysis makes use of the Run II data set recorded between April 2002 and May 2008 corresponding to an integrated luminosity of 2.8 fb^{-1}. A total of 107 $t\bar{t}$ candidate events with one electron and one muon in the final state are selected.

Applying the Matrix Element method to this data set, the top quark mass is measured to be

$$m_{\text{top}}^{\text{Run IIa}} = 170.6 \pm 6.1 \, (\text{stat.})_{-1.5}^{+2.1} \, (\text{syst.}) \, \text{GeV}$$

$$m_{\text{top}}^{\text{Run IIb}} = 174.1 \pm 4.4 \, (\text{stat.})_{-1.8}^{+2.5} \, (\text{syst.}) \, \text{GeV}$$

$$m_{\text{top}}^{\text{comb}} = 172.9 \pm 3.6 \, (\text{stat.}) \pm 2.3 \, (\text{syst.}) \, \text{GeV}.$$

Systematic uncertainties are discussed, and the results are interpreted within the Standard Model of particle physics. As the main systematic uncertainty on the top quark mass comes from the knowledge of the absolute jet energy scale, studies for a simultaneous measurement of the top quark mass and the b jet energy scale are performed. The prospects that such a simultaneous determination offer for future measurements of the top quark mass are sketched.

The outline of the present thesis is as follows. Chapter 2 gives a short summary of the experimental environment of the analysis, i.e., a description of the Tevatron collider and the DØ experiment. Chapter 3 focuses on the reconstruction of physics objects such as jets and leptons from the raw data of the detector. Additionally, the calibration of these objects is discussed. A short summary of top quark physics with a focus on production and decay of top quarks is given in Chap. 4. Moreover, the concept of mass and the relevance of the top quark mass for the Standard Model are discussed. A detailed description of the method and the improvements compared to earlier measurements [16] are presented in Chap. 5. The validation and calibration of the method, as well as the final measurement of the top quark mass and its uncertainties are the contents of Chap. 6. Detailed studies for an improved measurement of the top quark mass are discussed in Chap. 7. Chapter 8 summarizes and concludes this thesis.

References

1. Abe F et al (CDF collaboration) (1995) Observation of top quark production in $p\bar{p}$ collisions with the collider detector at Fermilab. Phys Rev L 74:2626
2. Aba S et al (DØ collaboration) (1995) Observation of the top quark. Phys Rev L 74:2632
3. The LEP collaborations, ALEPH, DELPHI, L3, OPAL, and the LEP Electroweak Working Group (1994) Combined preliminary data on Z parameters from the LEP experiments and constraints on the Standard Model. CERN-PPE/94-187
4. Hashimoto MTM et al (2001) Top mode standard model with extra dimensions. Phys Rev D 64:056003
5. Heinemeyer et al (2003) Physics impact of a precise determination of the top quark mass at an e^+e^- linear collider. JHEP 0309:075
6. Bigi I et al (1986) Production and decay properties of ultra-heavy quarks. Phys Lett B 181:157
7. Fiedler F (2007) Precision measurements of the top quark mass. Habilitationsschrift, Ludwig-Maximilians-Universität München
8. The Tevatron Electroweak Working Group for the CDF, DØ collaborations (2008) Combination of CDF and DØ results on the mass of the top quark. arXiv:hep-ex/0703034

9. The LEP collaborations, ALEPH, DELPHI, L3, OPAL, and the LEP Electroweak Working Group (2008) A combination of preliminary electroweak measurements and constraints on the Standard Model. CERN-PH-EP/2008-020

10. Abazov VM et al (DØ collaboration) (2004) A precision measurement of the mass of the top quark. Nature 429:638

11. Abazov VM et al (DØ collaboration) (2008) Precise measurement of the top quark mass from lepton+jets events. Phys Rev L 101:182001

12. Abulencia A et al (CDF collaboration) (2006) Top-quark mass measurement from dilepton events at CDF II. Phys Rev L 96:152002

13. Grohsjean A, Fiedler F (2008) Measurement of the top quark mass with the Matrix Element method in the dilepton channel. DØ note 5640

14. Abazov VM et al (DØ Collaboration) (2008) Measurement of the top quark mass in the electron–muon channel using the Matrix Element method with 2.8 fb^{-1}, DØ note 5743

15. Grohsjean A (2008) Measurements of the top quark mass in the dilepton decay channel at the DØ experiment. arXiv:hep-ph/08103711

16. Schieferdecker P (2005) Measurement of the top quark mass at DØ Run II with the Matrix Element method in the lepton+jets final state. Dissertationsschrift, Ludwig-Maximilians-Universität München

Chapter 2
Experimental Environment

The measurement presented in this thesis is based on a data set collected between April 2002 and May 2008 with the DØ experiment at the Fermilab. The collected data set corresponds to an integrated luminosity of 2.8 fb^{-1}. Besides the DØ detector, a second omni-purpose detector has been built at the Tevatron to study proton–antiproton collisions, CDF. During the so-called Run I period of the Tevatron accelerator (1991–1996), both CDF and DØ reported the discovery of the top quark in 1995.

The following chapter gives an overview of the experimental setup of this analysis. After a short description of the accelerator complex in Sect. 2.1, Sect. 2.2 describes the different components of the DØ detector, as well as the trigger system and the data acquisition.

2.1 The Tevatron Accelerator Complex

The Tevatron is a proton–antiproton collider with a circumference of 6.3 km and a center-of-mass energy of $\sqrt{s} = 1.96$ TeV. Up to the start of the LHC in 2009, it is the most energetic accelerator in the world, and the only place on Earth where top quarks can be produced directly. In the Run I period of the Tevatron, the center-of-mass energy was $\sqrt{s} = 1.8$ TeV, and both CDF and DØ collected an integrated luminosity of around 100 pb^{-1}. Starting in 1996, the accelerator was upgraded to achieve a center-of-mass energy of 1.96 TeV and a higher instantaneous luminosity. Nowadays peak luminosities are generally around 300 μb^{-1} s^{-1}.

Figure 2.1 shows the accelerator complex at the Fermilab, where the Tevatron itself is only the last one in a chain of seven accelerators. The most important parts of this complex are the following:

- *Cockroft–Walton preaccelerator* Hydrogen gas is ionized to create H$^-$ ions which are then accelerated to 750 keV.

A. Grohsjean, *Measurement of the Top Quark Mass in the Dilepton Final State Using the Matrix Element Method*, Springer Theses, DOI: 10.1007/978-3-642-14070-9_2, © Springer-Verlag Berlin Heidelberg 2010

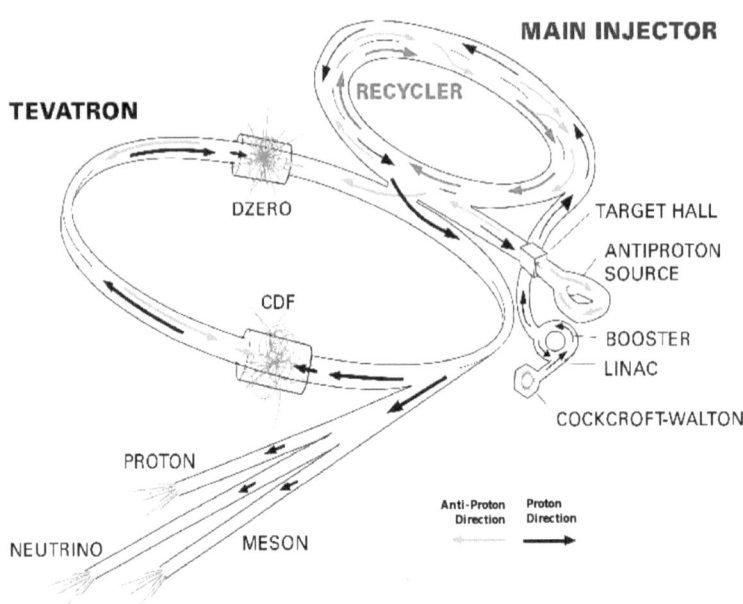

Fig. 2.1 Schematic overview of the Fermilab accelerator complex [1]

- *Linear accelerator (Linac)* The Linac is 150 m long and accelerates the H$^-$ ions to an energy of 400 MeV.
- *Booster* The Booster is the first synchrotron in the accelerator chain and has a circumference of about 475 m. A carbon foil strips the electrons from the H$^-$ ions at injection and leaves bare protons. The intensity of the proton beam is increased by injecting additional H$^-$ ions into the same orbit as the circulating ones. The protons are accelerated from 400 MeV to 8 GeV by a series of RF cavities.
- *Main Injector* The Main Injector is a synchrotron with a circumference of 3 km and two important key tasks. First of all, it accelerates protons from 8 to 120 GeV to produce antiprotons by directing the protons to a nickel target in the fixed target hall. Secondly, it accelerates both protons from the Booster and antiprotons from the Recycler to 150 GeV to inject them into the Tevatron.
- *Accumulator* In the Accumulator, a pulsed magnet separates the antiprotons from other particles and a lithium current lens is used to focus the beam. Additionally, the beam width is reduced by the so-called stochastic cooling. Here, pickup sensors sample the average transverse excursion for portions of each bunch, so that kicker magnets can apply correction forces to damp the antiproton beam on average.
- *Recycler* The Recycler is located in the same ring as the Main Injector. It recycles the antiprotons when the beam becomes poor after many collisions and integrates the cooled antiprotons in a stack.

- *Tevatron* In a last step, the Tevatron accelerates protons and antiprotons in opposite directions from 150 to 980 GeV. Superconducting magnets with a field of 4.2 T are used to keep the 36 bunches of protons and antiprotons on track. Each proton bunch contains 3×10^{11} particles, each antiproton bunch, 3×10^{10}. Finally they are brought to collision where the CDF and DØ detectors are built. The distance in time between two bunch crossings is 396 ns.

More information on the accelerator complex can be found in the design report [2].

2.2 The DØ Experiment

The DØ detector shown in Fig. 2.2 is one of the two omni-purpose detectors built at the Tevatron accelerator to study proton–antiproton collisions. It consists of three major subsystems: a tracking detector for vertex identification and momentum measurements, a uranium–liquid argon calorimeter to measure the energy of electromagnetic and hadronic showers, and a muon spectrometer to detect and measure muon momenta. After a short introduction of the coordinate system used at the DØ experiment, the most important aspects of the different subcomponents of the detector are discussed. A detailed description of the DØ detector is provided by the DØ collaboration [3].

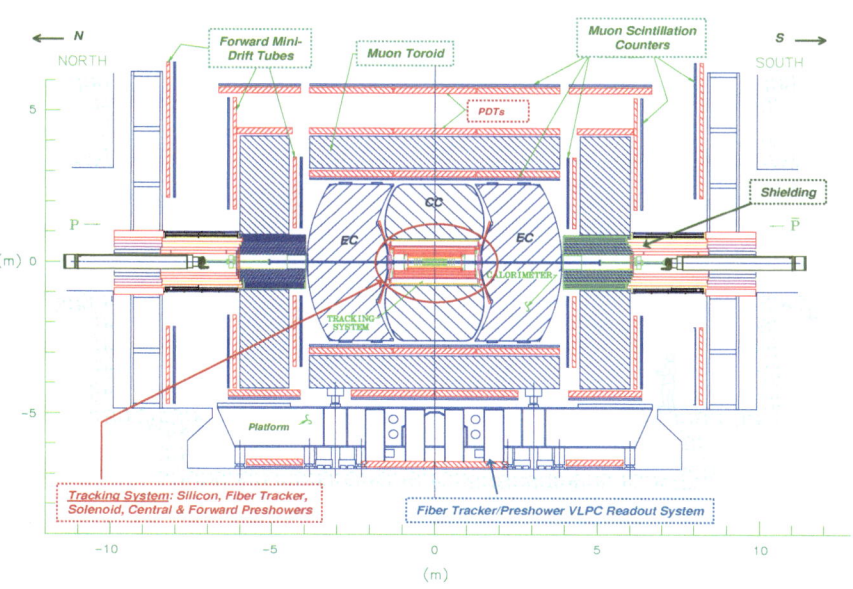

Fig. 2.2 Schematic profile of the DØ detector [4]

2.2.1 The Coordinate System

The right-handed coordinate system of the DØ detector is chosen such that its origin is at the nominal collision point. The x axis points to the center of the Tevatron ring, the y axis upwards, and the z axis along the direction of the proton beam. Alternatively to the polar angle θ, the pseudorapidity η is used. It is given by

$$\eta = -\ln\left(\tan\frac{\theta}{2}\right). \tag{2.1}$$

Consequently, the spatial separation between two particles is quantified in terms of

$$\Delta R = \sqrt{\Delta\eta^2 + \Delta\phi^2}, \tag{2.2}$$

where $\Delta\phi$ denotes the azimuthal angle.

2.2.2 The Tracking System

The barrel-shaped tracking system shown in Fig. 2.3 consists of the silicon microstrip tracker (SMT) and the central fiber tracker (CFT) within a solenoidal magnetic field of 2 T. First, it is designed to measure the momentum of charged particles via the curvature of their track within $|\eta| < 3$. Secondly, it separates electrons from pions, and detects secondary vertices for b quark identification. The primary interaction vertex resolution is about 35 μm and the momentum resolution in the central region about $\Delta p_T/p_T = 17\%$ for $p_T = 100$ GeV [5].

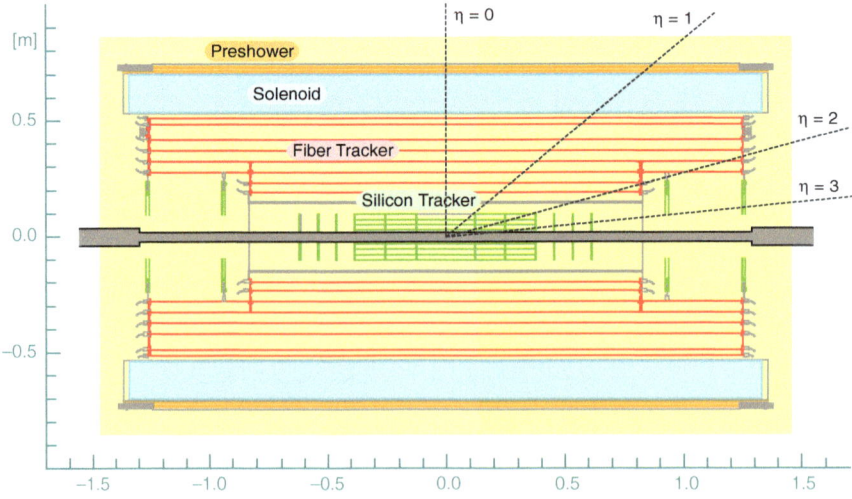

Fig. 2.3 Schematic drawing of the central tracking system [6]

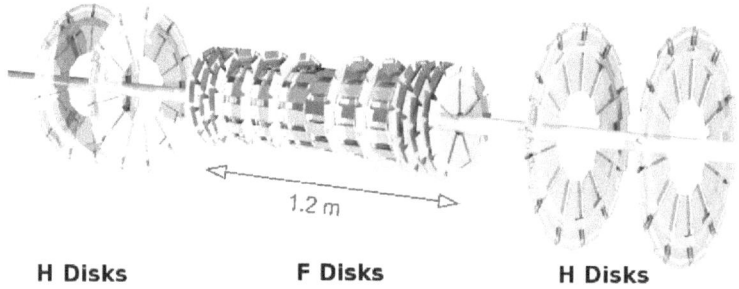

Fig. 2.4 Arrangement of the disks and barrels in the SMT [7]

2.2.2.1 Silicon Microstrip Tracker

The SMT shown in Fig. 2.4 uses reverse biased p–n junctions to detect particle tracks. It covers a range up to $|\eta| < 3$. Any passing charged particle causes ionization and electron-hole pairs are produced in the semiconductor. Reading out the charge deposition separately on every strip, the position of the particle can be measured with excellent resolution.

The SMT is composed of three subdetectors: the central barrels, the F disks and the H disks. The central barrels are made out of 5 (before 2005 only four) double layers of silicon detectors, and the 12 F disks, out of 12 double-sided wedge detectors. Six F disks are located in between the barrels with one attached to each end of the barrel detector, and six F disks, in a small distance from either end of the barrel detector. The two large-diameter H disks consisting of 24 wedges are mounted about 1 m from the interaction point. Each of the 24 wedges is constructed out of two back-to-back single-sided wedges. The barrels are used to measure the r–ϕ component of a particle, while the disks can measure both r–ϕ and r–z. The SMT provides a hit resolution of around 10 μm.

2.2.2.2 Central Fiber Tracker

The CFT surrounds the SMT with eight cylindrical layers of two fiber doublets, and provides a coverage up to $|\eta| < 1.6$. When a charged particle penetrates one of the fibers, the scintillator emits light via a rapid fluorescence decay. To increase the mean free path length of the light in the fiber, the fiber absorbs well at $\lambda = 340$ nm and emits at $\lambda = 530$ nm. The light is finally collected on one side of the fiber by a wave guide carrying the scintillation light to visible light photon counters for readout.

For each cylindrical layer, there is one doublet layer of fibers oriented parallel to the beam axis and one doublet layer enclosing an angle of ϕ with the beam axis, where ϕ alternately is $+3°$ and $-3°$ starting with $+3°$ from the innermost layer. To fill all gaps, each doublet is made of two layers with an offset of half a fiber width to each other. The CFT has a hit resolution of about 100 μm.

2.2.2.3 The Preshower Detectors

Outside the solenoid magnet, two additional tracking detectors are located: the central and the forward preshower detectors (CPS and FPS). The CPS is mounted on the solenoid and covers the range of $|\eta| < 1.3$, the FPS is on the inner surface of the end-calorimeter cryostat and covers the range of $1.5 < |\eta| < 2.5$. The lead absorbers of the preshower detectors convert electromagnetic particles (electrons and photons) into showers, and the shower energy is measured by several layers of scintillator strips. Besides the precise measurement of tracks in addition to the central tracker, the preshower detectors are used to correct the electromagnetic energy measurement of the central and the end calorimeters for losses in the solenoid and upstream material. In addition, they help to identify electrons and reject background.

2.2.3 The Calorimeter

The calorimeter shown in Fig. 2.5 consists of three parts embedded in three separate cryostats with a temperature of 90 K: the central calorimeter (CC) covering a range up to $|\eta| = 1$, and two end-cap calorimeters, one south of the interaction point (ECS) and one north (ECN) extending the covered range up to $|\eta| = 4$.

The main purpose of the calorimeter is the measurement of the energy of photons, electrons, and jets by inducing them to produce electromagnetic and hadronic showers. In addition, it helps to identify photons, electrons, jets, and muons, and it is used to measure the transverse energy balance in events.

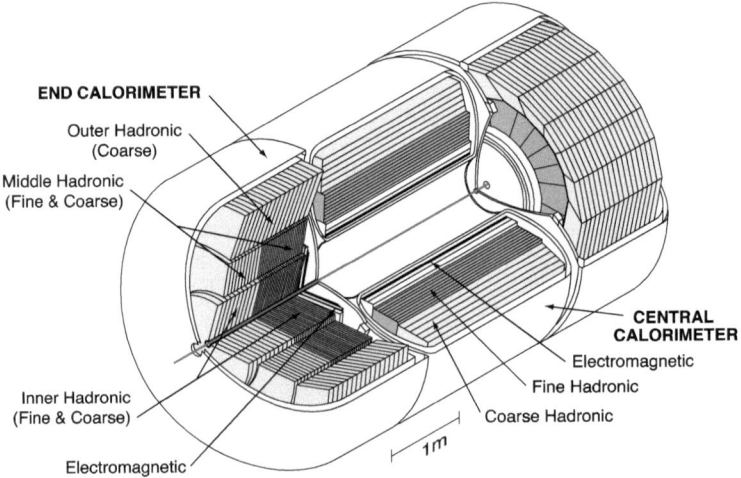

Fig. 2.5 Schematic drawing of the Run II DØ calorimeter [6]

Each of the three calorimeters contains an electromagnetic (EM) section closest to the interaction region followed by fine (FH) and coarse hadronic (CH) sections. The active medium of the calorimeter is always liquid argon, whereas the absorber in the electromagnetic section is depleted uranium, in the fine hadronic modules, uranium–niobium (2%) alloy, and in the coarse hadronic modules, copper (CC) or stainless steel (EC). The active medium is needed to sample and measure the energy of the shower through ionization, while the absorber causes the particles to shower.

The so-called inter-cryostat detector (ICD) is situated in between the CC and EC to compensate for the dead region between the cryostats. It consists of one layer of scintillating tiles mounted on the cryostat and is read out by photo-tubes that are connected by wavelength-shifting fibers.

The transverse size of the readout cells are comparable to the one of showers: 1–2 cm for electromagnetic showers and about 10 cm for hadronic showers. The granularity in η and ϕ is 0.1×0.1. The third layer of the EM modules, located at the electromagnetic-shower maximum, is segmented twice as finely to allow a more precise location of the electromagnetic-shower centroids.

The energy resolution of electromagnetic and hadronic objects in the calorimeter can be parameterized as

$$\frac{\Delta E}{E} = \sqrt{\frac{S^2}{E\,(\text{GeV})} + \frac{N^2}{E^2\,(\text{GeV})^2} + C^2}, \tag{2.3}$$

where S describes fluctuations in the energy deposition, N, instrumental effects like uranium noise, and C, uncertainties in the calibration. These parameters are measured from data and listed in Table 2.1.

2.2.4 The Muon Spectrometer

The muon system surrounds all other detector components and consists of two parts: the central muon system covering the region up to $|\eta| = 1$, and the forward muon system covering the region of $1 < |\eta| < 2$. A toroidal magnet of 1.8 T allows a separate measurement of the muon momentum in the muon spectrometer. Gas filled tubes are used to collect and to measure the ionization created by the passing muons.

For a central muon with a transverse momentum of 40 GeV, the momentum resolution is measured to be $\Delta p_T/p_T = (9.6 \pm 0.3)\%$ [10]. The resolution is

Table 2.1 Parameters for the energy resolution in the calorimeter [8, 9]

	C	$S\,(\sqrt{\text{GeV}})$	$N\,(\text{GeV})$
Electron, photon	0.041	0.15	0.29
Jet	0.036	1.05	2.13

limited by multiple scattering at low momentum and the individual hit resolution at high momentum. Nevertheless, it allows a cleaner matching with the central tracks, rejection of π and K in-flight decays, and yields an improved momentum resolution at high muon momenta.

2.2.4.1 The Central Muon System

The central muon system consists of three layers A, B, and C of proportional drift tubes (PDTs), see Fig. 2.6. While layer A is located between the cryostat and the toroidal magnet, layer B and C are outside the toroidal magnet. Since the drift time of the PDTs (750 ns) is longer than the bunch crossing time of the Tevatron, two additional layers of scintillators shown in Fig. 2.7 are used to trigger muon events: the A-ϕ counters between the calorimeter and layer A, and the cosmic caps mounted outside layer C. The A-ϕ counters are also used to reject cosmic muons and scattered particles from the calorimeter, the cosmic caps are used to reject cosmic muons. Due to the support structure of the DØ detector, the bottom part of the detector is only partly covered with scintillator counters. The resolution of the PDTs is about 1 mm.

2.2.4.2 The Forward Muon System

The forward muon system consists of three layers A, B, and C of mini drift tubes (MDTs) and scintillator counters, see Fig. 2.6. The MDTs have been newly built for Run II. Here, also layer A is between the cryostat and the toroidal magnet, while layer B and C are outside the magnet. Though the drift time of the MDTs is

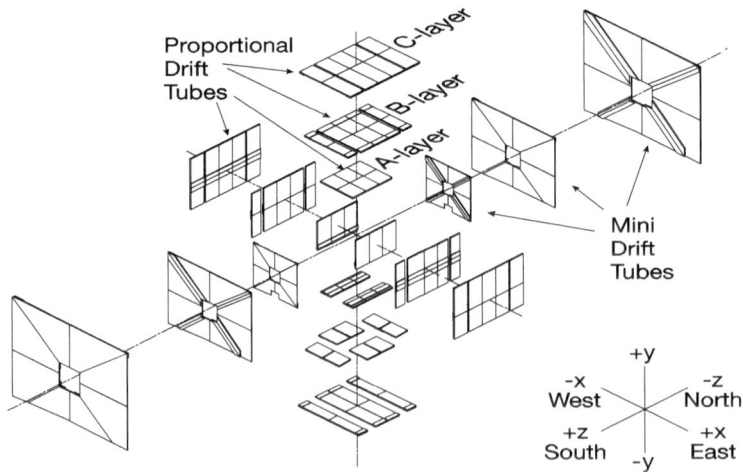

Fig. 2.6 Arrangement of the PDTs and MDTs in the muon system [3]

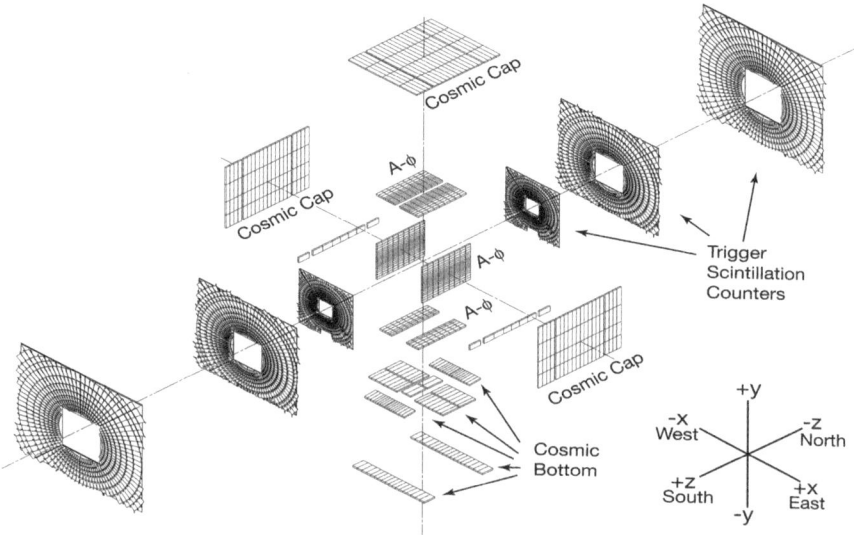

Fig. 2.7 Arrangement of the scintillation detectors in the muon system [3]

only 60 ns, scintillation counters are also used to trigger muon events and to reject cosmic muons and other sources of background. The forward muon system has additional shielding around the beam pipe to reduce trigger rates and the aging of the detector by beam halo. The resolution of the MDTs is about 0.7 mm.

2.2.5 The Luminosity Monitors

The instantaneous luminosity is given by

$$\mathscr{L} = \frac{1}{\sigma_{p\bar{p}}^{\text{eff}}} \frac{\mathrm{d}N_{p\bar{p}}}{\mathrm{d}t}, \qquad (2.4)$$

where $\sigma_{p\bar{p}}^{\text{eff}}$ denotes the effective inelastic $p\bar{p}$ cross section measured by the luminosity monitors taking detector acceptance and efficiency into account. The luminosity monitors are two hodoscopes built from plastic scintillators mounted on the cryostats of the end-cap calorimeters. They cover a pseudorapidity of $2.7 < |\eta| < 4.4$. In order to distinguish between $p\bar{p}$ interactions and beam halo, the z coordinate of the interaction vertex is calculated from the difference in time-of-flight between the two luminosity monitors. Particles from inelastic $p\bar{p}$ collisions have a smaller time-of-flight difference than particles from the beam halo. The overall estimated error on the luminosity is 6.5%.

The integrated luminosity is calculated in so-called luminosity blocks. Each luminosity block is indexed by a luminosity block number. After each run or store

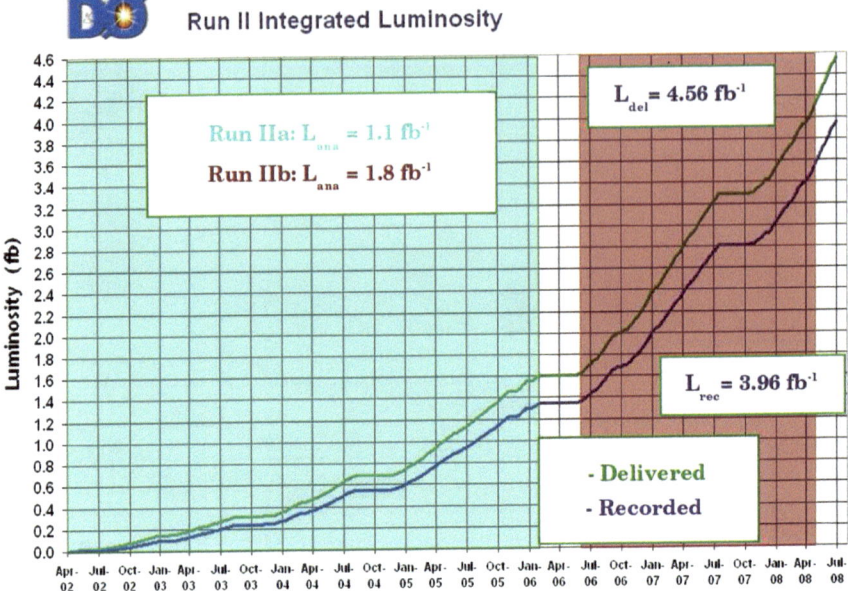

Fig. 2.8 Integrated luminosity delivered to and recorded by the DØ experiment in Run II until July 2008

transition the number monotonically increases. The time period is chosen such that for each luminosity block the integrated luminosity is about the same.

Figure 2.8 shows the integrated luminosity delivered to and recorded by the DØ experiment. The average data taking efficiency is about 89% and up to now, 4 fb^{-1} are stored on tape by the DØ experiment. The measurement presented in this analysis uses the full reconstructed data set of 2.8 fb^{-1}. The data set is split in two periods, Run IIa from April 2002 to February 2006 with 1.1 fb^{-1}, and Run IIb from June 2006 to May 2008 with 1.7 fb^{-1}. It has to be noticed that the Run I data set is not included as the center-of-mass energy was increased between Run I and Run II, and a measurement of the charged track momentum was not possible during Run I.

2.2.6 The Trigger System and Data Acquisition

The trigger system is designed to reduce the event rate delivered by the Tevatron to an event rate that can be written to tape and is practical to analyze. Most of the events produced in $p\bar{p}$ collisions are multijet events, while the production cross section of massive bosons, top quarks, or particles beyond the Standard Model is extremely small. To select these events, the trigger system is designed to select

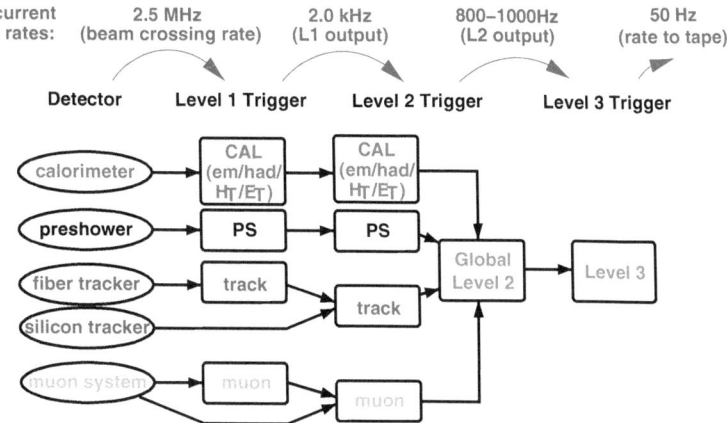

Fig. 2.9 Outline of the DØ trigger system

events with high transverse momentum jets or leptons. The DØ trigger system consists of three levels: Level 1 (L1), a pure hardware trigger using electronic signals from the detector, Level 2 (L2), using both electronic signals and reconstructed physics objects, and Level 3 (L3), a pure software trigger, using only reconstructed objects. A diagram of the trigger system is shown in Fig. 2.9.

2.2.6.1 Level 1 Trigger

The L1 trigger reduces the event rate from 2.5 MHz to about 2 kHz in about 4.2 μs. It consists of a luminosity trigger, a calorimeter trigger, a central-track trigger and a muon-system trigger. The luminosity system provides indication that a collision occurred, and the tracking system reconstructs tracks and stores seed track candidates to be used by other trigger units. The calorimeter performs a fast summation of electromagnetic and hadronic towers to look for towers that exceed a certain threshold. The muon triggers require coincidence between signals in layer A and either B or C.

2.2.6.2 Level 2 Trigger

The L2 trigger reduces the event rate to 1 kHz and takes about 100 μs to make a decision. The system consists of two stages. In the preprocessor stage, the information from the different subsystems is used to reconstruct physics objects; in the global stage, the information across the subsystems is used to build physics objects. The preprocessor stage consists of the L2 calorimeter trigger, the L2 preshower trigger, the L2 muon-system trigger, the L2 SMT trigger, and the L2 central-track trigger.

2.2.6.3 Level 3 Trigger

The L3 trigger finally reduces the rate to about 50 Hz. This software trigger uses a simplified event reconstruction algorithm and is fully based on reconstructed physics objects.

2.2.6.4 The Data Acquisition System

The transport of the selected data events from the readout crates to the farm nodes is done with the data acquisition (DAQ) system. The triggering and the data acquisition are controlled by a coordination program called COOR. The readout of the luminosity monitors is done with a standalone DAQ system.

References

1. The Tevatron accelerator division. http://www-bd.fnal.gov/public/relativity.html.
2. The Tevatron accelerator division (1988) Proton–antiproton collider upgrade, FERMILAB-DESIGN-1988-01
3. Abazov VM et al (DØ collaboration) (2006) The upgraded DØ detector. Nucl Instrum Meth A 338:463
4. The DØ collaboration. http://www-d0.fnal.gov/Run2Physics/displays/presentations/#gallery
5. Ellison J (2001) The DØ detector upgrade and physics program. DØ note 3830
6. The DØ collaboration. http://www-d0.fnal.gov/Run2Physics/top/top_public_web_pages/top_dzero_detector.html
7. The DØ collaboration. http://www-d0.fnal.gov/Run2Physics/WWW/drawings.htm
8. Chaplin D et al (2005) Measurement of $Z \rightarrow ee$ and $W \rightarrow ev$ production cross sections using one tight electron. DØ note 4897
9. Royon C et al (2007) Jet pT resolution using v7.1 JES for p17 data. DØ note 5381
10. Calfayan P et al (2006) Muon identification certification for p17 data. DØ note 5157

Chapter 3
Event Reconstruction and Simulation

Before triggered $t\bar{t}$ candidate events can be analyzed, physics objects, such as jets, leptons, muons and missing transverse energy, need to be reconstructed from the raw data output of the detector. The measured energies and momenta of the decay products then allow an extraction of the top quark mass from the selected data sample. In order to study the properties of the data sample and to calibrate the method for the mass measurement, Monte Carlo simulated events are used. In the following sections, the reconstruction of the physics objects is described with a focus on objects used in this analysis: electrons (3.3), muons (3.4), jets (3.6), and missing transverse energy (3.8). The calibration of the charged leptons is discussed in Sect. 3.5, the calibration of the jets, in Sect. 3.7. At the beginning of this chapter, a brief description of the basic objects, such as tracks (3.1) and vertices (3.2), is given. After the reconstruction of the physics objects, the data quality criteria used at the DØ experiment are summarized in Sect. 3.9. The chapter ends with a discussion of the event simulation used to verify and calibrate the measurement technique.

3.1 Tracks

The reconstruction of tracks is one of the most important steps in the reconstruction of physical objects. Vertices are identified by clustering tracks, and the identification of electrons and muons requires a central track match, while photons are required to have no central track. In addition, the reconstruction of jets can be significantly improved by combining calorimeter and tracking information.

Charged particles passing the tracking detector leave hits in the detector that can be reconstructed as a track. The track finding is done in two steps. First of all, a set of track candidates is made using two different algorithms [1, 2]. Finally, the combined candidate list is passed to a track fitter based on a Kalman Filter

A. Grohsjean, *Measurement of the Top Quark Mass in the Dilepton Final State Using the Matrix Element Method*, Springer Theses, DOI: 10.1007/978-3-642-14070-9_3, © Springer-Verlag Berlin Heidelberg 2010

algorithm [3]. Removing ambiguities and refitting all tracks with the Kalman Filter, the final track parameters are calculated with proper uncertainties.

3.2 Vertices

Particles traveling through the detector cause tracks that originate in one vertex. The so-called primary vertex (PV) represents the point of the hard interaction of the proton and antiproton, while the secondary vertices (SV) correspond to decaying heavy hadrons. The exact knowledge of vertices is important to determine precisely the direction of any calorimeter object.

3.2.1 Primary Vertex

The primary vertex is defined as the interaction point of the proton and antiproton. A reliable reconstruction of the primary vertex is needed to distinguish objects from the hard interaction and the inelastic scattering. The primary vertex is reconstructed in three steps using the Adaptive Vertex Fitting [4]. First of all, tracks with a minimum transverse momentum of 0.5 GeV and at least two hits in the acceptance region of the SMT are selected. Then, all tracks of a cluster within 2 cm around the beam axis are constrained to a common vertex, and the vertex position, as well as all track parameters are recalculated. The fitting is repeated until the χ^2 distance of all tracks to the fitted vertex converges. Finally, the hard scattering vertex is selected as the vertex with the lowest probability to arise from a collision without hard interaction, i.e. with low transverse momentum [5].

3.2.2 Secondary Vertex

Heavy hadrons travel up to several millimeters in the detector before they decay. The decay results in a secondary vertex displaced from the primary vertex. Thus, secondary vertices can be used to detect heavy hadrons and to identify jets from the decay of a b hadron. The reconstruction of secondary vertices is based on a Kalman Filter [6] and consists of four steps. First of all, jets are reconstructed based on tracker information only, then a list of tracks from these jets is selected, the vertex finding is performed, and the secondary vertices are finally selected.

3.3 Electrons

Electrons are identified as narrow clusters in the electromagnetic calorimeter, where all towers in a cone of radius $\Delta R = 0.2$ around the most energetic seed are numbered among the cluster. The cluster is required to have a minimum transverse energy of 1.5 GeV.

Since electrons lose most of their energy E_{tot} in the electromagnetic calorimeter, the cluster is required to deposit 90% of its energy in the electromagnetic layers E_{EM}. In addition, the cluster has to be isolated, i.e. no significant additional activity in a cone of radius $R = 0.4$ is allowed,

$$\frac{E_{tot}(R<0.4) - E_{EM}(R<0.2)}{E_{EM}(R<0.2)} < 0.15. \tag{3.1}$$

To distinguish electrons from other particles, the shower of each electron candidate is compared to average distributions from simulated electron showers, and a χ^2 as a measure of compatibility is assigned. An electron candidate must fulfill $\chi^2 < 50$. As electrons are charged particles, at least one reconstructed track with a transverse momentum of at least 5 GeV is required to point to the cluster of the electron candidate with

$$|\Delta\eta| \times |\Delta\Phi| < 0.05 \times 0.05. \tag{3.2}$$

The major remaining background arises from photons which happen to overlap with a track from a nearby charged particle. To discriminate against this background, an electron likelihood $L_{EM,7}$ is calculated using seven different variables [7]. Electrons are required to have $L_{EM,7} \geq 0.85$.

Because of the poor energy resolution of electrons reconstructed in the inter-croystat region, electron candidates with a pseudorapidity between 1.1 and 1.5 are rejected. To suppress multiple scattering background, electrons are also not allowed to exceed a pseudorapidity of 2.5.

Comparing the reconstructed mass peak in $Z \rightarrow ee$ data and Monte Carlo simulated events, a better resolution and a shifted peak has been observed in Monte Carlo. Hence the electron energies in Monte Carlo simulated events have to be smeared and scaled accordingly to reproduce the data [8].

3.4 Muons

The reconstruction of muons uses information from both the muon system and the tracking detector. The former provides an unambiguous identification of the muon, the latter a precise measurement of its momentum.

To discriminate muons against misidentified hadrons, the central track of the muon in the tracking chamber extrapolated to the muon system is required to match the reconstructed track in the muon detector. In addition, muon candidates are required to lie in the fiducial region of the muon system, i.e. the pseudo-rapidity must be less than 2. According to the number of wire and scintillator hits in the muon system and the quality of the matched track, muons are classified into different muon and track qualities, which are described by Cal-fayan et al. [9].

To reduce the number of non-isolated muons from semimuonic heavy-hadron decays in jets, a distance of at least 0.5 in R between the reconstructed muon and all jets in the event is required. For tight isolated muons two additional isolation criteria are applied. One variable is computed by summing up the energies of the reconstructed tracks, the other variable is derived from energy deposited in the calorimeter. For background muons, the size of either of these sums is correlated with the muon energy, while for signal, it is not. Therefore, scaling the sums by the transverse muon momentum generates variables that tend to be higher for background than for signal. The track-based variable is computed as

$$\varepsilon_{\text{halo}}^{\text{trk}} = \frac{1}{p_T^\mu} \sum_{\Delta R < 0.5} p_T^{\text{trk}}, \tag{3.3}$$

where the track matched to the muon is excluded from the sum. Similarly, the calorimeter-based isolation is defined as

$$\varepsilon_{\text{halo}}^{\text{cal}} = \frac{1}{p_T^\mu} \sum_{0.1 < \Delta R < 0.4} E_T^{\text{cell}}, \tag{3.4}$$

where the sum is over individual calorimeter cells. Tight isolated muons are required to have values of both $\varepsilon_{\text{halo}}^{\text{trk}}$ and $\varepsilon_{\text{halo}}^{\text{cal}}$ that are less than 0.15.

Cosmic ray muons are rejected by requiring a timing window of 10 ns for all hits in the scintillator starting at the expected time of arrival from the hard interaction point. If any hit in the three layers lies outside this window, a muon is tagged as cosmic.

As for electrons, the simulated resolution of muons is different from the measured one and an additional smearing is applied to the simulated data [9].

3.5 Calibration of Charged Leptons

The calibration of electrons and muons is done using $Z \rightarrow ee$ or $Z \rightarrow \mu\mu$ events, as well as $c\bar{c}$ and $b\bar{b}$ resonance decays. These events have a clear back-to-back signature in the detector and can be identified with low backgrounds. The energy of the electrons is measured in the electromagnetic calorimeter, the momentum of the muons, in the central tracking chamber, and thus both can be calibrated.

The total efficiency to reconstruct an electron or a muon is the product of several contributions: the trigger efficiency, the efficiency of track identification, and the efficiency of applied quality cuts. All individual efficiencies are measured with the so-called tag-and-probe method using $Z \rightarrow ee$ or $Z \rightarrow \mu\mu$ events. Here, the criterion to be studied is only applied to one of the leptons. The fraction of selected events where the second lepton also fulfills this

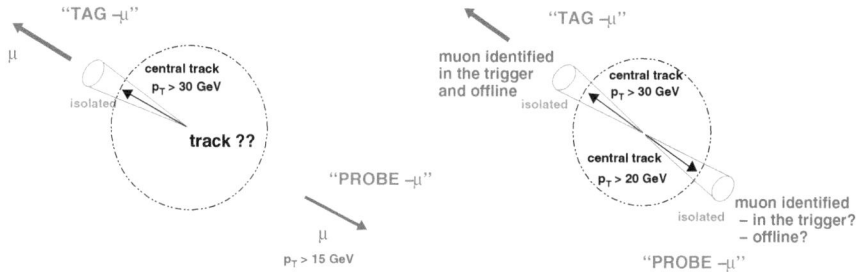

Fig. 3.1 Illustration of the tag-and-probe method [10]; *left*: measurement of the tracking efficiency; *right*: identification of a muon in the trigger system and offline

requirement then yields a measurement of the efficiency. An illustration of this procedure is shown in Fig. 3.1.

3.6 Jets

Quarks and gluons hadronize immediately and form showers of mesons and baryons in the hadronic calorimeter. The more energetic the parton is, the tighter the final hadrons are concentrated around the direction of the parton. At the DØ experiment, a cone algorithm with a radius of $\Delta R = 0.5$ is used to reconstruct jets [11]. The transverse energy of a jet relative to the beam line is defined under the assumption that it originates at the primary vertex position.

The reconstruction of jets is done in three steps. First of all, preclusters are formed from calorimeter towers. Then, all preclusters are associated iteratively with a jet until the energy weighted center of the so-called proto jet is stable. Finally, the proto jets get either split or merged according to their energy overlap. All reconstructed jets with a transverse energy of less than 6 GeV are removed.

To suppress the fraction of fake jets, different quality criteria are applied. In order to distinguish jets from electrons, the electromagnetic energy fraction of a jet is required to be less than 95%. A minimum electromagnetic fraction of 5% is required to discriminate against noise. In addition, the ratio of the energy from the triggered L1 towers to the energy from the precision readout must be at least 0.4 taking into account that noise does not appear simultaneously in two independent readout chains. Noise dominated jets are removed by requiring the energy fraction in the coarse hadronic calorimeter to be less than 40%. To suppress jets clustered around single cells that fired erroneously, two additional cuts are applied. First, the hot fraction, i.e. the energy ratio of the highest to the next-to-highest calorimeter cell assigned to a jet, must be less than 10%. Secondly, it is not allowed that a single cell contains 90% of the total jet energy. Far forward jets with a pseudorapidity of more than 2.5 are eliminated to reduce background from beam remnants. The cut values quoted above refer to jets measured in the central region of the detector. A detailed list of the cut values for the different detector regions is given by A. Harel [12].

Jets with a distance of less than 0.5 in R to a high energetic electromagnetic object are rejected.

3.7 Jet Energy Scale

For any measurement at a $p\bar{p}$ collider, it is mandatory to have a good knowledge of the jet energy scale (JES) and the jet resolution. Their determination is more difficult than that for a charged lepton and the uncertainties on the jet energy scale and resolution are the dominant source of systematic uncertainty in the measurement of the top quark mass.

3.7.1 Overall Jet Energy Scale

The goal of the jet energy scale is to correct the energy of the calorimeter jet back to the one of the stable-particle jet before any interaction with the detector took place, see Fig. 3.2. After this is done for both light and b jets, an additional correction is applied to jets from bottom quarks. The correction from calorimeter to particle jets involves five steps.

Fig. 3.2 Evolution from a parton to a calorimeter jet [13]

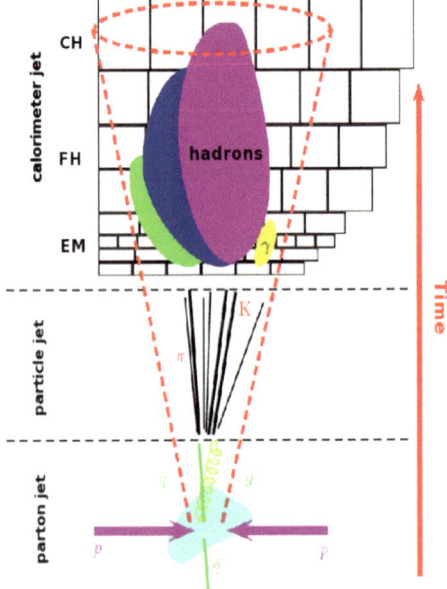

3.7.1.1 Energy Offset E_O

E_O represents an offset to the jet energy that includes contributions from noise coming from electronics and radioactive decays of the uranium absorber, additional $p\bar{p}$ interactions, and previous crossings. The energy contributed by the spectator quarks to the hard-scattering process, the so-called underlying event, is not included in the offset as it already affects jets at the particle level. The offset energy depends on the cone size of the jet, its pseudorapidity with respect to the primary vertex, the number of reconstructed vertices and the instantaneous luminosity.

The offset is estimated from the energy density in minimum-bias and zero-bias events. Minimum-bias events are triggered by the luminosity monitors signaling the presence of a potential inelastic proton-antiproton collision. Their energy density is a measure for the effect from multiple proton-antiproton collisions. To eliminate contributions from the spectator quarks, the difference of the energy density in events with more than one primary vertex and exactly one primary vertex is used. Zero-bias events are triggered at every bunch crossing with the explicit requirement of no vertex in the event. Their energy density is a measure of contributions from pile-up, electronics, and uranium noise.

The energy offset of a jet with cone radius 0.5 is shown for different primary vertices as a functions of its pseudorapidity in Fig. 3.3.

3.7.1.2 Absolute Calorimeter Response R

The largest correction is the one that accounts for the calorimeter response of the jet. It is measured in $\gamma + $ jet events where the detector response of the photon is known with little uncertainty from $Z{\rightarrow}ee$ events. The so-called Missing E_T

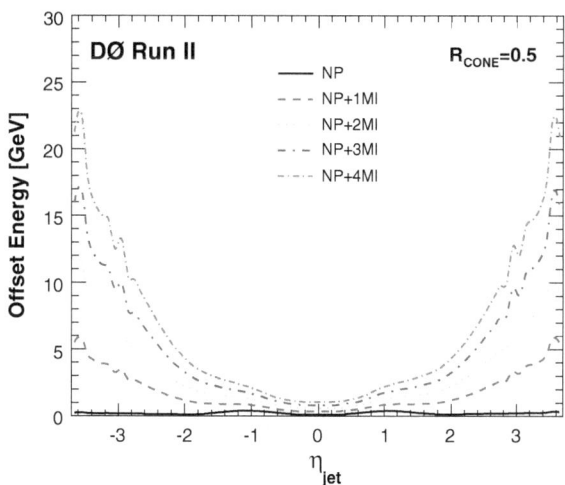

Fig. 3.3 Offset energy for a cone radius of 0.5 as a function of the jet pseudorapidity [13]. The *different lines* show the prediction for the offset energy from noise and pile up (NP) and multiple interaction (MI)

Projection Fraction method (MPF) allows a measurement of the calorimeter response from the p_T imbalance in the event. In the calorimeter, the response of the photon and the hadronic recoil, corresponding to the jet in the ideal case of one photon and exactly one jet, might be different. This results in a transverse momentum imbalance $\vec{\not{p}}_T$ measured in the calorimeter

$$R_\gamma \vec{p}_T^{\,\gamma} + R_{\text{had}}\vec{p}_T^{\,\text{had}} = -\vec{\not{p}}_T, \tag{3.5}$$

where $R_\gamma \vec{p}_T^{\,\gamma}$ corresponds to the measured photon momentum, and $R_{\text{had}}\vec{p}_T^{\,\text{had}}$, to the jet momentum. Taking into account that the true hadronic recoil and the photon momentum balance each other, Eq. 3.5 can be written as

$$\frac{R_{\text{had}}}{R_\gamma} = 1 + \frac{\vec{\not{p}}_T \cdot \vec{p}_T^{\,\gamma}}{(\vec{p}_T^{\,\gamma})^2}. \tag{3.6}$$

Thus, the response of the hadronic recoil relative to the response of the photon can be estimated from the photon momentum and the projection of the imbalance of the transverse momentum $\vec{\not{p}}_T$ onto the direction of the photon in the transverse plane.

The absolute response measurement is performed using events with one single photon candidate in the central region of the detector, and exactly one jet with an absolute pseudorapidity of less than 0.4. The photon and the jet are required to be back-to-back, so $\Delta\phi(\gamma, \text{jet}) > 3.0$. Since the energy resolution of jets is broader than the one of photons, the dependency on the true jet energy is estimated from the transverse energy of the photon by

$$E' = p_T^\gamma \cosh(\eta_{\text{jet}}), \tag{3.7}$$

where η_{jet} is measured with respect to the reconstructed primary vertex. The estimator E' is strongly correlated with the particle-level-jet energy since it is calculated using the transverse momentum of the photon and the direction of the jet which are measured with good precision. The energy dependence of the jet response is described by a quadratic logarithm function.

$$R(E') = p_0 + p_1 \log(E'/100(\text{GeV})) + p_2 \log(E'/100(\text{GeV})). \tag{3.8}$$

Due to uncertainties in the detector simulation, the calorimeter response is measured separately in data and Monte Carlo simulated events. The response for jets with a cone radius of 0.5 in $\gamma + $ jet data events is shown in Fig. 3.4. Extrapolating the fit to very high jet energies, the response uncertainty is about 2%.

3.7.1.3 Relative Calorimeter Response F_η

While the central part and the end caps of the calorimeter are almost uniform, the region in between is not as well instrumented which causes a non-uniformity in the response as a function of the pseudorapidity. The relative response correction aims

Fig. 3.4 Measured response in γ + jet data events for a jet cone size of 0.5 [13]

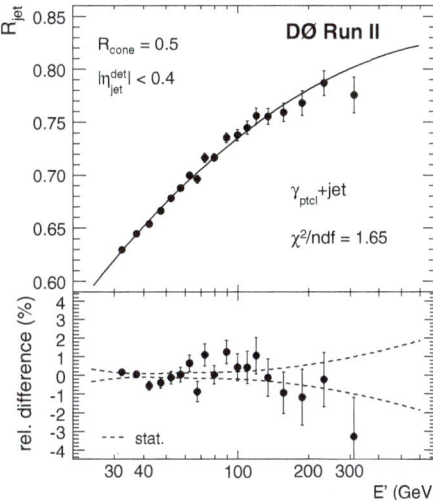

to calibrate forward jets with respect to the central ones, and gives the final response correction in combination with the absolute scale.

The relative response is measured as the absolute response in γ + jet data and Monte Carlo events using the Missing E_T Projection Fraction method. The tag object, i.e. the photon, is required to have a pseudorapidity of less than 0.4, and the response of the probe jet is determined using a fine binning of the pseudorapidity (mostly about 0.1). The relative response as a function of the pseudorapidity for different values of the energy estimator E' defined in Eq. 3.7 are depicted in Fig. 3.5. As expected, the correction is largest in the region of $1.0 < |\eta| < 1.5$. Due to the additional material in front of the calorimeter, the response to lowly energetic particles is much smaller than in Run I.

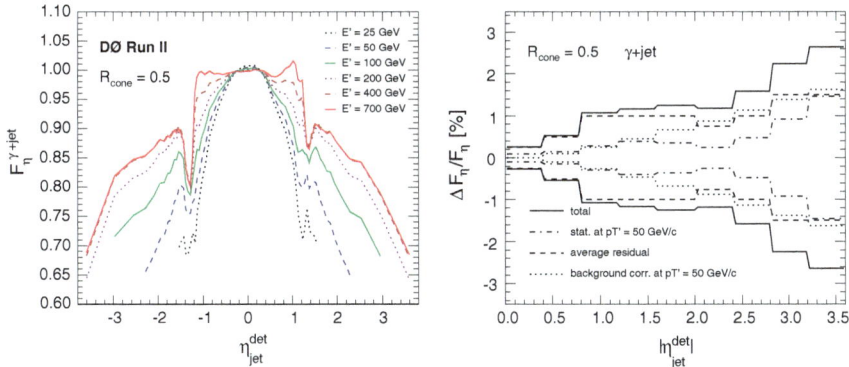

Fig. 3.5 *Left*: measured calorimeter response as a function of the pseudorapidity for different values of the jet energy estimator E' in γ + jet data events; *right*: corresponding uncertainties [13]

The estimated uncertainties on the relative response are shown on the right side of Fig. 3.5. The dominant uncertainty arises from imperfections on the energy-dependent parameterization and η interpolation (average residuals). In the very forward region, there is also a significant contamination from background.

3.7.1.4 Showering S

The showering correction represents a correction of the fraction of the energy deposited outside (inside) the jet cone from particles (not) belonging to the jet as a result of the development of showers in the calorimeter and the finite size of calorimeter cells. Typically, the net correction is smaller than unity. It strongly depends on the cone size of the jet and its pseudorapidity with respect to the primary vertex. The showering correction is measured separately in data and Monte Carlo simulated γ + jet events. To reduce the impact of multiple interactions, exactly one primary vertex is required.

First, the jet energy density profile from calorimeter towers is measured as a function of the radial distance to the jet axis. Then, templates of various contributions to the jet profile from Monte Carlo simulated events are fitted to the measured jet profile adjusting the relative size of the specific profile source. Finally, the showering correction is determined as the ratio of the visible energy in the jet cone to the visible energy of all particles from the particle jet. In Monte Carlo simulated events, the showering correction also can be determined using information about the energy deposit of each particle in the different calorimeter cells. The results are in good agreement with the template-based method.

The size of the showering correction is shown in Fig. 3.6. It is much bigger for forward than for central jets, as the solid angle of a jet decreases with increasing jet rapidity making the detector showering effects more important.

Fig. 3.6 Showering correction for jets with a cone size of 0.5 in data as a function of the jet energy estimator E' [13]

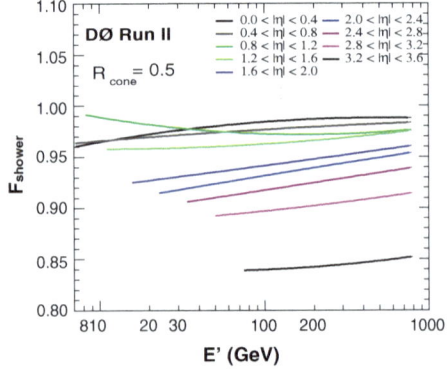

3.7.1.5 Remaining Bias k_{bias}

To achieve an accuracy on the order of 2%, an additional factor is applied to correct for the major remaining biases. Due to the additional energy in a jet, the low energy cells can pass the zero-suppression threshold, so that the offset measured in zero-bias and minimum-bias events does not entirely correspond to the offset in a jet. A similar bias is introduced in the response measurement because the photon is much more compact than the jet. The two effects are studied in Monte Carlo simulated events by comparing the response and offset for the same jet in samples with and without zero-bias overlay. It turned out that both effects cancel each other partially, and the residual correction is small but not negligible.

Moreover, additional activity in the event or the splitting and merging of jets can bias the measurement of the jet response. These effects have been studied using Monte Carlo simulated events where individual particle energy depositions in different calorimeter cells can be used to compute the true jet response to which the response from the Missing E_T Projection Fraction method can be compared.

Taking all corrections into account, the true energy of the particle jet, E_{corr}, can be calculated from the measured jet energy, E_{meas}, as

$$E_{\text{corr}} = \frac{E_{\text{meas}} - E_0}{R \cdot F_\eta \cdot S} \cdot k_{\text{bias}}, \qquad (3.9)$$

where E_0, R, F_η, S and k_{bias} are the corrections described above.

The relative uncertainty on the jet energy scale in data as a function of the energy estimator E' is shown in Fig. 3.7 for jets with a cone size of 0.5 and three different pseudorapidity bins of 0.0–0.5, 1.0–1.5, and 2.0–2.5. Besides the total relative uncertainty, the contributions from the different subcorrections are displayed, where the relative and the absolute response corrections have been lumped together. The relative uncertainty is mainly caused by the uncertainty on the response, especially at low and high transverse energies. The dominant source here is the uncertainty on the photon energy scale coming from the absolute scale uncertainty from the $Z \rightarrow ee$ mass peak and the relative electron-to-photon scale uncertainty. At low transverse energies and in the very forward region, the background contamination in $\gamma + $ jet events plays an important role, at high transverse energies the main problem is the limited statistics to constrain the high energy extrapolation. All technical details and studies regarding the Run II JES are given by Juste et al. [14].

3.7.2 Bottom Quark Jet Energy Scale

Jets from bottom quarks and light-flavor jets, i.e. all jets not initiated by a bottom quark, show different frequencies in the production of hadrons and their momentum spectra. Thus, the ratio of electromagnetic and hadronic energy is

Fig. 3.7 Relative uncertainty on the jet energy scale for jets with a cone radius of 0.5 in data as a function of the uncorrected transverse energy and three different pseudorapidity bins; *top*: $0.0 < |\eta_{jet}| < 0.5$; *middle*: $1.0 < |\eta_{jet}| < 1.5$; *bottom*: $2.0 < |\eta_{jet}| < 2.5$. Besides the total fractional uncertainty, the one from the response, the showering, and the offset are shown [13]

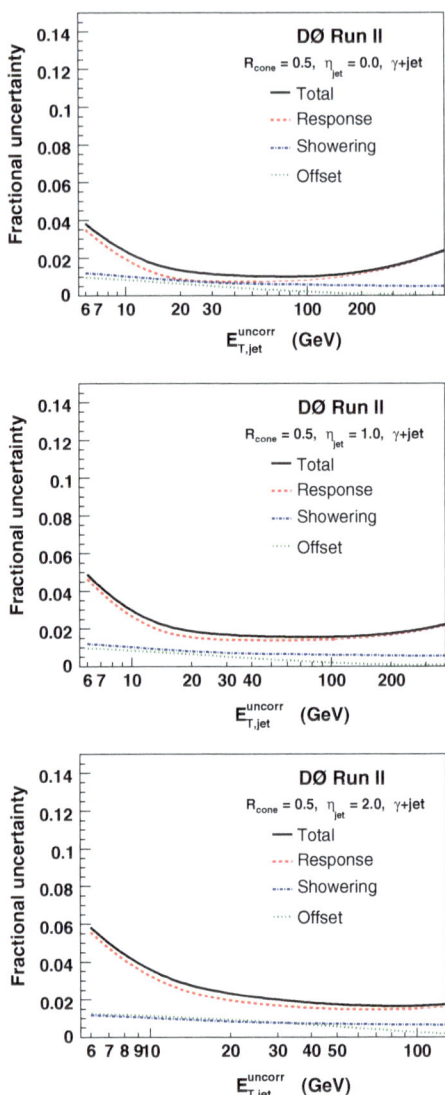

different for both kinds of jets leading to a different response. In addition, *b* jets may contain a semileptonic heavy-hadron decay causing a different response in case this decay is not detected. An explicit correction of the jet energy is applied when the charged lepton is identified inside the jet. Due to the large mass of a bottom quark, the showering for light-flavor and *b* jets is also different. So far, there is no certified jet energy scale for jets initiated from *b* quarks at the DØ experiment. Thus, the overall jet energy scale is used for both light-flavor and *b* jets and the additional systematic uncertainty is assigned to the reconstructed energy of *b* jets.

3.7.3 Jet Energy Resolution

The jet energy resolution is measured using γ + jet and Z + jet events [15]. Its parameterization is given in Eq. 2.3. Comparing the energy resolution in measured and simulated events, the resolution of the simulated jets needs to be smeared and shifted to reproduce the data. In addition the reconstruction and identification efficiencies are different for data and Monte Carlo simulated events, such that some of the simulated jets need to be discarded. All this is done in the so-called Jet Shifting, Smearing and Removal (JSSR) [16].

3.8 Missing Transverse Energy $\not{E}_T$

Since neutrinos escape the detector without any interaction, they can only be identified indirectly by the imbalance of the transverse momentum of the event. The missing transverse momentum is calculated from the vector sum of all cal-orimeter cells[1] which pass the T42 algorithm [17]. The missing momentum vector is corrected for the jet energy scale and the scale of any charged lepton in the event. In addition, it is corrected for the energy loss of muons in the calorimeter through ionization that is substantially smaller than the muon momentum mea-sured in the tracking chamber.

The unclustered transverse energy is defined as the magnitude of the vector sum of the transverse momenta of all calorimeter objects that are not assigned to a jet or a charged lepton.

3.9 Data Quality

In order to ensure a good data quality, the following selection criteria are applied. A whole run or luminosity block, see Sect. 2.2.5, is excluded if one of the subdetector systems is declared as bad during data taking. In addition, a list of bad luminosity blocks is created based on certain calorimeter information that is checked per luminosity block. Events belonging to a bad luminosity block are discarded.

In addition, there are calorimeter noise patterns, for which single events are discarded. The noise patterns include the so-called noon noise, which is caused by external noise, coherent noise, where for all crates a synchronous shift in the pedestals occur, and ring of fire, where a ring of energy in the transverse plane

---

[1] Due to the large noise in the coarse hadronic layers, cells here are only considered if clustered into a reconstructed jet.

occurs based on grounding problems. Events are tagged with empty crates in case one of the crates could not be read out.

3.10 Monte Carlo Simulation

Monte Carlo (MC) simulation is one of the key tools in experimental high energy physics. It is used for several purposes in the present analysis. First of all, Monte Carlo events are used to optimize the selection of $t\bar{t}$ events, and to check the sample composition of the selected data sample. Secondly, simulated events are used to calibrate the method for measuring the top quark mass, and to compare the expected uncertainties to the measured ones.

The simulation of Monte Carlo events is based on the leading-order matrix element of the considered process. The MC generator ALPGEN [18] is used to generate the hard parton-scattering process for the signal process $t\bar{t}$ and the main background process $(Z \rightarrow \tau\tau)$ + ets. PYTHIA [19] is used to simulate the background processes involving two weak vector bosons (WW + jets, WZ + jets). The simulation of the hard-scattering process is interfaced with PYTHIA to model initial- and final-state gluon radiation. To avoid an overlap of the phase-space regions covered by the hard-gluon radiation and the gluon emission from the matrix-element calculation, a matching procedure [20] has been used. PYTHIA is also used to model the fragmentation and hadronization. They are interfaced with EVTGEN [21] and TAUOLA [22] to describe the heavy hadron and τ lepton decays.

Afterwards, the MC simulated events are processed with a detailed simulation of the DØ detector based on GEANT [23], and the same reconstruction and selection criteria as for the data are applied.

Depending on the instantaneous luminosity, more than one $p\bar{p}$ collision may take place in the same bunch crossing. To simulate this effect, minimum-bias events are measured, and added to the simulated events. In a similar way, pile-up of signals from previous collisions is recorded with a random trigger and added.

As discussed in Sects. 3.3, 3.4, and 3.7.3, the energy and momentum resolutions of the simulated events are slightly different from the measured data. Therefore, an additional smearing and shifting is applied to the generated distributions to be consistent with the recorded data.

References

1. Khanov A (2000) HTF: histogramming method for finding tracks. DØ note 3778
2. Borissov G http://www-d0.fnal.gov/atwork/adm/d0_private/2003-02-28/adm_talk.ps
3. Greenlee H (2003) The DØ Kalman track fit. DØ note 4303
4. Schwartzman A et al (2005) Primary vertex reconstruction by means of Adaptive Vertex Fitting. DØ note 4918
5. Schwartzman A et al (2002) Probabilistic primary vertex selection. DØ note 4042

6. Schwartzman A et al (2001) Secondary vertex reconstruction using the Kalman filter. DØ note 3908
7. Wang L et al (2006) Electron likelihood efficiency in p17. DØ note 5114
8. Gris P (2007) Electron smearing studies with Run IIa data. DØ note 5400
9. Calfayan P et al (2006) Muon identification certification for p17 data. DØ note 5157
10. Nurse E et al (2004) Measurement of the cross section for inclusive Z production in dimuon final states at $\sqrt{s} = 1.96$ TeV DØ note 4573
11. Blazey GC et al (2000) Proceedings of the workshop QCD and weak boson physics in Run II. FERMILAB-PUB-00-297 47
12. Harel A (2005) Jet ID optimization. DØ note 4919
13. Juste A et al http://www-d0.fnal.gov/phys_id/jes/public_RunIIa
14. Juste A et al (2007) Jet energy scale determination at DØ Run II. DØ note 5382
15. Royon C et al (2007) Jet pT resolution using v7.1 JES for p17 data. DØ note 5381
16. Grivaz JF et al (2008) SSR for p17. DØ note 5609
17. Bernardi G et al (2004) Improvements from the T42 algorithm on calorimeter object reconstruction. DØ note 4335
18. Mangano ML et al (2003) ALPGEN, a generator for hard multiparton processes in hadronic collisions. JHEP 307:1
19. Sjöstrand T et al (2006) PYTHIA 6.4 physics and manual. JHEP 605:26
20. Begel M et al (2006) Determination of weighting factors for ALPGEN Monte Carlo signal and background samples. DØ note 5016
21. Lange DJ, et al. (2001) The EVTGEN particle decay simulation package. Nucl Instrum Meth A 462:152
22. Decker R et al (1993) The decay library TAUOLA. Comput Phys Commun 76:361
23. Brun R et al (1993) GEANT detector description and simulation tool. CERN Programming Library Long Writeup W5013

Chapter 4
The Top Quark and the Concept of Mass

The present chapter describes the fundamental elements of this thesis: the top quark and the concept of mass. After a short review of top quark properties in Sect. 4.1, its production in Sect. 4.2 and decay in Sect. 4.3 are described. Afterwards, different concepts of mass and the relevance of the top quark mass are discussed in Sects. 4.4 and 4.5. The chapter ends with a short overview of different mass measurement concepts.

4.1 The Top Quark within the Standard Model

When the bottom quark was discovered in 1977, the Standard Model of particle physics [1, 2] predicted the top quark as the weak-isospin partner of the bottom quark. Indirect evidence for the top quark became compelling over the years, and constraints from electroweak data [3] pointed exactly to the range where the top quark was finally discovered 18 years later. The discovery of the top quark in 1995 by the CDF [4] and DØ experiments [5] was a great success for the Standard Model, but compared to the other quarks and leptons far less is known about the top. The strong interaction is mostly studied in the pair production of top quarks, the weak interaction, in its decay and in the production of single top quarks. An overview of the particles in the Standard Model as well as their masses is given in Fig. 4.1.

Four fundamental parameters of the Standard Model are directly associated with the top quark: its mass and the three CKM matrix elements involving top. With the current world average of 172.4 ± 0.7(stat.) ± 1.0(syst.) GeV [6], the top quark is the heaviest of the observed fundamental particles and it is often discussed to play a special role in the electroweak symmetry breaking. Besides the large mass of the top, the top quark is also unique due to the theoretical expected lifetime of 5×10^{-25} s [7], where the experimental bound is $c\tau_{\text{top}} < 52$ μm at

A. Grohsjean, *Measurement of the Top Quark Mass in the Dilepton Final State Using the Matrix Element Method*, Springer Theses, DOI: 10.1007/978-3-642-14070-9_4, © Springer-Verlag Berlin Heidelberg 2010

95% CL [8]. The top quark is predicted to decay so quickly that it cannot hadronize and the spin of the top is not depolarized by chromomagnetic interactions within bound states. Thus, it is possible to measure observables which depend directly on the spin of the top quark giving an excellent test of the Standard Model.

The electric charge of the top quark is predicted by the Standard Model to be $Q = +2/3$, but is not yet measured. Thus, the fact that the discovered particle at the Tevatron could be an exotic quark with charge $Q = -4/3$ decaying into a negatively charged W boson and bottom quark is only excluded at the 92% CL [9].

In the Standard Model, the top quark is predicted to have the same V–A charged current weak interaction as the other fermions. In the limit of a massless bottom quark, the b quark has to be left-handed, so that the W boson cannot be right-handed assuming angular momentum conservation and the top quark to be a spin-1/2 particle. Since the coupling of the top quark to a longitudinal (zero helicity) W boson is similar to its Yukawa coupling, the branching ratio is predicted to be about 70%. The by-now most precise measurement yields a branching ratio of -0.03 ± 0.07 for the coupling of the top quark to a right-handed W boson, and 0.66 ± 0.16 for the longitudinal. So both measurements are in perfect agreement with the Standard Model.

Fig. 4.1 Overview of the leptons and quarks within the Standard Model as well as the gauge-mediating bosons and the yet-unobserved Higgs boson [10]. The particles are ordered according to their mass. Two particles in the Standard Model are massless: the photon and the gluon

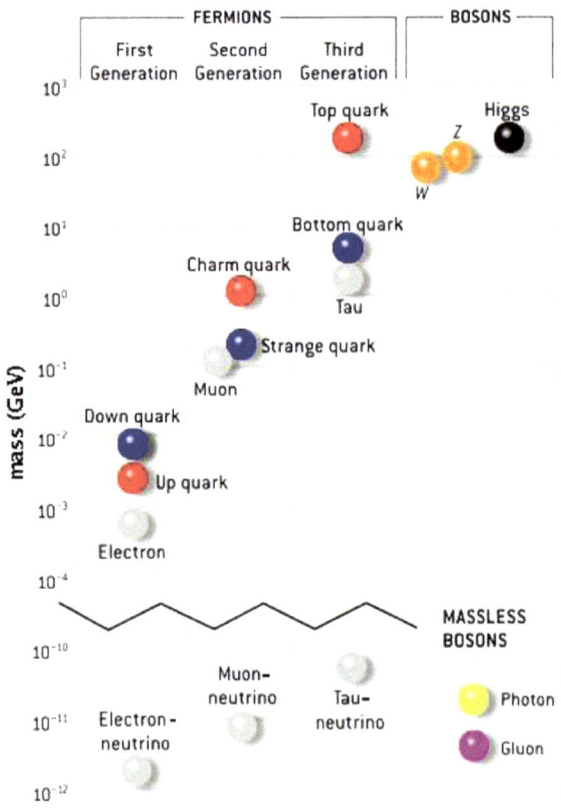

4.2 Top Quark Production

Until the start of the LHC in 2009, the Tevatron collider is the only place on Earth where top quarks can be produced directly. Due to the high energies, the production can be described by quantum chromodynamics and factorized in three different parts: the modeling of the partons of the incoming hadrons, the hard-scattering process of two partons, and the hadronization of the final-state particles. A schematic illustration of these parts is given in Fig. 4.2. The constituents of the colliding hadrons are modeled using the so-called parton distribution functions (PDFs). The PDF $f_{\mathrm{PDF}}(\varepsilon, \mu_F^2)$ describes the probability to find a gluon or quark of given flavor with longitudinal momentum fraction ε within the colliding parton. The factorization scale μ_F^2 denotes the energy region where the PDFs are probed and finally allows the separation of the parton modeling and the hard-scattering process. Since the PDFs cannot be calculated directly, they have to be measured in fits to experimental data [11]. As an example, the CTEQ6L1 parameterization at a scale of $\mu_F = 170$ GeV used in the present analysis is shown in Fig. 4.3.

At the Tevatron collider, top quark pairs are produced via the strong interaction, while single top quarks are produced via the electroweak one. The leading-order Feynman diagrams for both production processes are shown in Figs. 4.4 and 4.5.

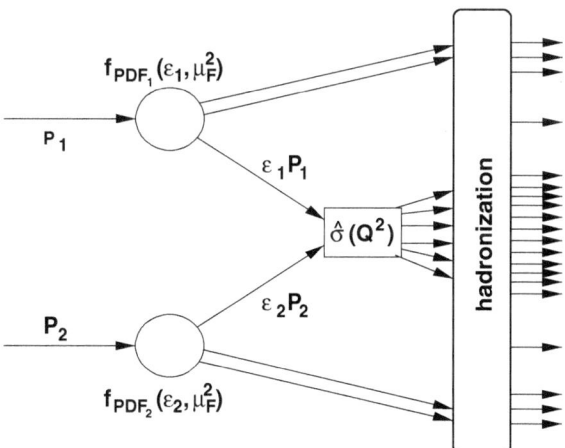

Fig. 4.2 Schematic illustration of a hard-scattering process between two hadrons. Due to the high energies the internal structure of the colliding hadrons can be resolved, and the hard interaction takes place between the two partons with momentum fractions $\varepsilon_1 P_1$ and $\varepsilon_2 P_2$ from the incoming hadrons with momentum P_1 and P_2. The partonic cross section $\hat{\sigma}$ can be calculated perturbatively using the factorization μ_F and the renormalization scale μ_R. The formation of the final-state particles cannot be calculated in perturbation theory, but again is independent of the hard interaction

Fig. 4.3 CTEQ6L1 parame-
trization of the parton distribution functions depending
on the momentum fraction ε
of the parton. The factorization scale μ_F is chosen to be
170 GeV

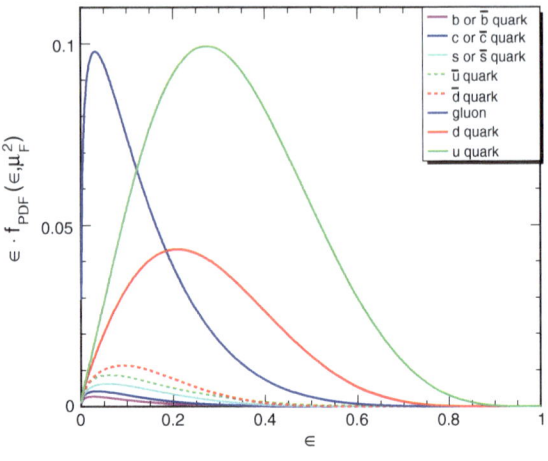

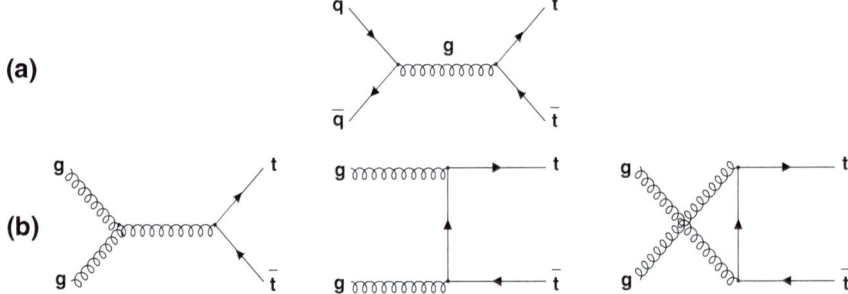

Fig. 4.4 The leading-order Feynman diagrams for the production of top quark pairs at a hadron
collider [12]. At the Tevatron, 85% of the top quark pairs are produced in quark–antiquark
annihilation (**a**), while only 15% are produced in gluon fusion (**b**)

Fig. 4.5 The leading-order
Feynman diagrams for the
production of single top
quarks at a hadron collider
[12]. At the Tevatron, 28.5%
are produced in the s-channel,
64% in the t-channel and only
7.5% in association with a W
boson

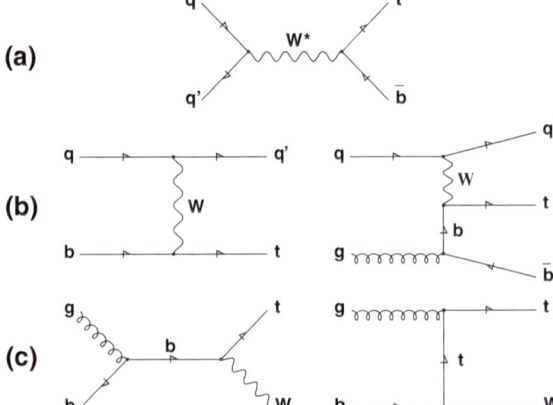

The $t\bar{t}$ production cross section is given by

$$\sigma = \sum_{p_1, p_2} \int d\varepsilon_1 d\varepsilon_2 f_{PDF}(\varepsilon_1, \mu_F^2) f_{PDF}(\varepsilon_2, \mu_F^2) \hat{\sigma}\left(\varepsilon_1, \varepsilon_2, \alpha_S(\mu_F^2), \frac{Q^2}{\mu_R^2}\right), \quad (4.1)$$

where $\sum_{p_1, p_2}$ denotes the sum over all contributing initial-state parton species, ε_1, ε_2, the momentum fractions of the interacting partons, and $\hat{\sigma}\left(\varepsilon_1, \varepsilon_2, \alpha_S(\mu_F^2), \frac{Q^2}{\mu_R^2}\right)$, the hard-scattering cross section that is convoluted with the parton distribution functions, $f_{PDF}(\varepsilon_1, \mu_F^2)$, and $f_{PDF}(\varepsilon_2, \mu_F^2)$. Besides the energy fractions, ε_1 and ε_2, the hard-scattering cross section, $\hat{\sigma}$, also depends on the strong coupling constant, α_S, at the factorization scale, μ_F^2, the ratio of the transferred momentum squared, Q^2, and the renormalization scale, μ_R^2. The latter one is similar to a cut-off parameter introduced to renormalize divergences from higher-order contributions in the production of top quarks. The factorization scale, μ_F^2, and the renormalization scale, μ_R^2, are often chosen to be the same. According to N. Kidionakis et al. [13], the theoretical cross section for $t\bar{t}$ production at a center-of-mass energy of 1.96 TeV is calculated to be 7.91 ± 0.50 pb. At the Tevatron, 85% of the top quark pairs are produced in $q\bar{q}$ annihilation, while only 15% are produced in gluon fusion. At the LHC however, with pp collisions at a center-of-mass energy of 14 TeV, the gluon-induced processes dominate with 90%, since the gluon PDFs are much larger at small ε than the quark PDFs. The total cross section for $t\bar{t}$ production at the LHC is two orders of magnitude larger than at the Tevatron.

A first evidence for single top quark production was observed end of 2007 by the DØ collaboration. Nevertheless, single top quarks are not yet used to measure the top quark mass. The cross section of single top production is calculated for the three different channels shown in Fig. 4.5 to be 0.98 ± 0.04 pb in the s-channel, 2.2 ± 0.1 pb in the t-channel, and 0.26 ± 0.06 pb in the W associated production [14].

4.3 Top Quark Decay

For an assumed top quark mass of 170 GeV, the width of the top quark is calculated to be 1.5 GeV, which corresponds to a lifetime of about 5×10^{-25} s. Thus, the lifetime of the top quark is much shorter than the time needed for hadronization and the top quark decays before it can form bound states. The Standard Model predicts the top quark to decay almost exclusively to a W boson and a bottom quark via the weak interaction, where the branching ratio is larger than 0.998 at the 95% confidence level [15]. Thus, the possible final states of the top quark can be classified according to the decay of the W boson, which decays with branching fraction 2/3 into two quarks, and 1/3 into a charged lepton and the corresponding neutrino. The different branching fractions of $t\bar{t}$ pairs are shown in Fig. 4.6. Since τ leptons are difficult to reconstruct and the muonic or electronic decay of a τ lepton leads to 2 additional neutrinos, final states with

Top Pair Decay Channels   **Top Pair Branching Fractions**

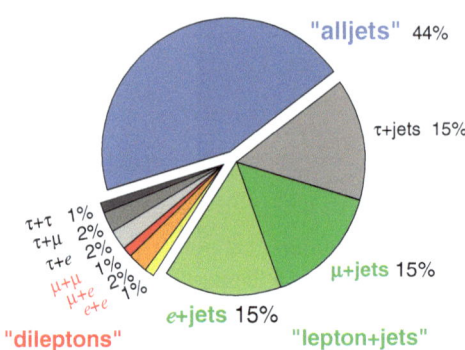

Fig. 4.6 Top pair decay channels and branching fractions [16]

tauonic W decay are not included, so that the expression lepton in this thesis usually refers to an electron or a muon. Commonly, three different decay channel are distinguished: the dilepton channel, the lepton + jets channel, and the alljets channel.

4.3.1 Dilepton Channel

With only 5%, the dilepton channel is the channel with the smallest contribution. Its topology is described by two highly energetic b jets, as the b quarks hadronize and shower in the detector, two isolated leptons, and missing transverse energy $\not{E}_T$ from the two undetected neutrinos. Thus, the reconstruction of the final-state particles in the dilepton channel is underconstrained. The reconstruction of the neutrinos requires six constraints, where only five can be obtained assuming the top and antitop masses to be equal, the event to be balanced in the transverse plane, and using the known mass of the two W bosons. Thus, the kinematics can only be solved if the top quark mass is assumed or additional information is used, such as the relative probability for different configurations of final-state particle momenta. On the other hand, the dilepton channel is the cleanest channel of all. The largest physics background is from Z boson production in association with jets. In the dielectron or dimuon channel, this background can be reduced by requiring the invariant mass of the two leptons to be different from the Z mass. In the decay channel where one W boson decays into an electron and the other one into a muon, the main background comes from the tauonic Z decay, so that the requirement of highly energetic objects in the final state helps to discriminate against the Z background. The main instrumental background in this channel comes from events in which a jet or photon is misidentified as an electron and from the production of heavy hadrons which decay into leptons passing the isolation requirements.

4.3.2 Lepton + Jets Channel

The lepton + jets channel has a fraction of 29% and is characterized by two highly energetic b jets, two highly energetic jets from the hadronic W decay, and one isolated lepton and missing transverse energy from the corresponding neutrino. Here, the neutrino momenta can be reconstructed using e.g. the missing transverse energy and the mass of the leptonically decaying W boson. However, the reconstruction of the top quark mass is not straightforward in this channel, since the association of measured jets with final-state quarks is not known. The number of possible combination can be reduced using b jet identification. The main source of background are events with a leptonically decaying W boson produced in association with jets. In addition, multijet events, where one electron fakes a jet, contaminate the sample. The usage of b jet identification helps to discriminate against the large fraction of background events where most of the jets are not b jets. The lepton + jets channel is the so-called golden channel. It allows for the most precise measurement of the top quark mass so far.

4.3.3 Alljets Channel

The alljets channel has the largest fraction of all channels with about 46%. Here, the topology is given by six highly energetic jets where two of these jets come from the b quarks. Even though all momenta of the final-state particles are well known, the reconstruction of the top quark mass in this channel is not trivial due to the 90 possible different jet-quark assignments[1].

In addition, this channel suffers from the huge multijet background, which cannot be easily described with Monte Carlo simulated events.

4.4 The Concept of Mass

Isaac Newton gave one of the first scientific definitions of the expression mass in 1687 in his landmark Principia [17]: "The quantity of matter is the measure of the same, arising from its density and bulk conjointly."

However, it is clear that this definition does not hold for a point-like, unstable, colored particle such as the top quark. The mass of the top is only defined in the scope of a given theory, in which it appears as a parameter. To determine this mass, the predictions of the model need to be compared to the measurements. In

---

[1] The number of possible combinations, $6! = 720$, is reduced by a factor of 2^3 as the reconstructed top quark mass is invariant under the exchange of the two jets combined to a W boson and the three jets to a top quark.

the following, two different ways of defining the top quark mass will be discussed. More details are given by the Particle Data Group [15].

- $\overline{MS}$ *Scheme* After renormalization, the Lagrangian of the QCD with the particle mass parameters m gives finite values for physical quantities such as the scattering amplitude. Thus, these mass parameters depend on the renormalization scheme and the dimensionful scale parameter μ used in this scheme. The most common renormalization scheme for QCD perturbation theory is the so-called Modified Minimal Subtraction ($\overline{MS}$) scheme [18].
- *Pole Mass* For an observable particle such as an electron or muon, the position of the pole in the propagator defines the mass of this particle. In perturbative QCD, this definition can also be used as a definition of quark masses [19]. However, because of confinement, i.e. non-perturbative effects in QCD, the full quark propagator does not have a pole, and the pole mass cannot be used to arbitrarily high accuracy. An uncertainty of order λ_{QCD}/m_{top} remains.

The relation between the $\overline{MS}$ mass and the pole mass is known up to three loops [15].

Since Monte Carlo simulated events are used to derive the calibration curve for the measurement, the measurement relies on the definition used in the Monte Carlo generator. In the present analysis, ALPGEN is used to simulate $t\bar{t}$ events. Here, the decaying top quark is described by a Breit–Wigner resonance with

$$f(p^2) \propto \frac{p^2}{\pi} \frac{m\Gamma}{(p^2 - m^2)^2 + (m\Gamma)^2}, \tag{4.2}$$

where p denotes the four-momentum, m, the mass and Γ, the decay width that is assumed to be constant. Thus, the measured mass can be interpreted as the mass parameter in the Monte Carlo generator which is approximately a pole mass.

4.5 Relevance of the Top Quark Mass

The precise knowledge of the top quark mass is of special interest as it allows the determination of parameters within the Standard Model and beyond.

As an example, the top and the W masses offer to infer the mass of the yet-unobserved Higgs boson. The propagator of the W boson receives contributions from loop diagrams, where massive particles contribute most. Figure 4.7 shows the lowest-order diagram for the W boson, (a) the contribution from the top quark, and (b) and (c), the one from the Higgs boson. The corrections that arise from these diagrams are quadratic in the top quark mass, but only logarithmic in the Higgs boson mass. Thus, precise measurements of both the top quark and the W boson masses constrain the Standard Model Higgs boson mass.

Figure 4.8 shows the agreement between the direct measurements of the W and top masses from LEP2 and the Tevatron experiments in blue, and the indirect

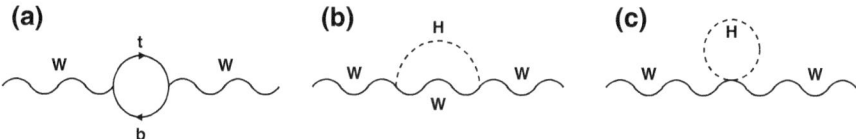

Fig. 4.7 Next-to-leading-order loop corrections to the mass of the W boson: from the top quark (**a**) and the Higgs boson (**b, c**) [12]

Fig. 4.8 Comparison between direct measurements of the top and W mass (*solid line*) and indirect constraints (*dashed line*), together with the Standard Model prediction of the relation between the top and W mass for various assumed Higgs masses [20]

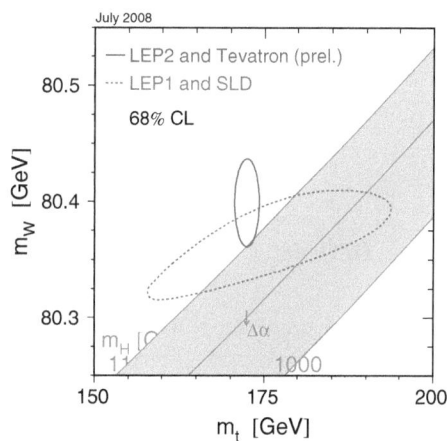

constraints from a Standard Model fit using LEP1 and SLD data in red. The relation between the top and W mass as a function of the Higgs mass is shown in green, where the left boundary corresponds to the LEP exclusion limit of 114 GeV and the right to an upper theoretical limit of 1,000 GeV. The direct measurements of the top quark and the W boson mass predict the Higgs to be light.

Figure 4.9 shows a combined fit of all precision data measurements to the Standard Model prediction as a function of the Higgs mass. The light blue band around the fit shows an estimate of the uncertainty from higher-order corrections that were not included in the calculation. Including the theoretical uncertainties, the 95% CL upper limit on the Standard Model Higgs mass is 154 GeV [21]. It shifts to 185 GeV when taking the lower limit of 114.4 GeV from LEP into account. With the first exclusion at 170 GeV by the Tevatron experiments this summer [22], the interesting phase of the search for the Standard Model Higgs boson has begun.

4.6 Measurement Methods of the Top Quark Mass

The mass of the top quark can be determined with three fundamentally different methods.

Fig. 4.9 χ^2 of a fit to the Standard Model parameters as a function of the Higgs mass [21]. The *light blue band* reflects the uncertainties from higher-order corrections. The yellow area corresponds to the Higgs mass range excluded by the LEP experiments

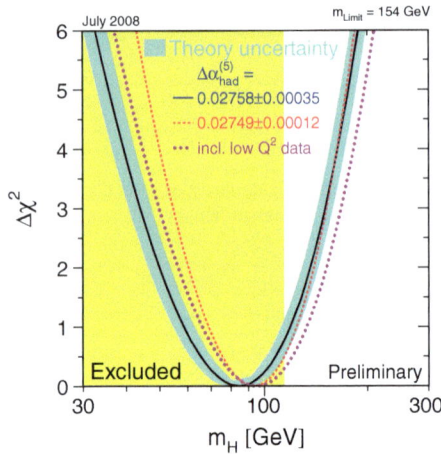

4.6.1 Indirect Constraints

Indirect constraints from the Standard Model parameters can be used to determine the top quark mass by comparing predictions from the Standard Model as a function of the top mass to the measured data. Indeed, the fits from electroweak data pointed exactly to the range where the top quark was finally discovered in 1995.

4.6.2 Reconstruction of Decay Products

The invariant mass of a particle is given by $m = \sqrt{E^2 - \vec{p}^2}$. Thus, energy and momentum conservation allows a measurement of the top quark mass by measuring the four-momentum vectors of the top decay products. Due to the fact that e.g. the neutrino momenta cannot be measured directly, different techniques are used to reconstruct the invariant top mass. An overview of these techniques is given by F. Fiedler [12].

4.6.3 Measurements of the $t\bar{t}$ Cross Section

In principle, there are two different approaches for this kind of measurement. The first one uses a likelihood that is built from the theoretically predicted and the measured $t\bar{t}$ cross sections as a function of the top mass [23]. The top mass is then given by the minimum of the likelihood function integrated over the cross section. The preliminary result of this measurement using the Run IIa data set is shown in

Fig. 4.10 Measured $t\bar{t}$ production cross section from the combined semileptonic and dileptonic decay channel as a function of the top quark mass. The comparison with the theoretical NLO + NNLL calculations from Moch and Uwer yields a top mass of $169.6^{+5.4}_{-5.5}$ GeV, while the comparison with the NLO + NLL calculations from Cacciari et al. yields $167.8^{+5.7}_{-5.7}$ GeV. The PDFs are parametrized with CTEQ6.6

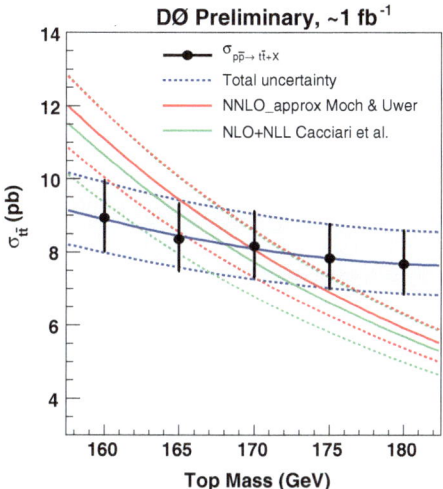

Fig. 4.10. The comparison with the theoretical NLO + NNLL calculations from S. Moch et al. [24] yields a top mass of $169.6^{+5.4}_{-5.5}$ GeV, while the comparison with the NLO + NLL calculations from M. Cacciari et al. [25] yields $167.8^{+5.7}_{-5.7}$ GeV. While direct measurements rely on the not-very-well-defined renormalization scheme of the Monte Carlo generators, used to simulate events and their behavior, this method makes use of the measured cross section and the fully inclusive theoretical calculations in higher-order QCD including soft gluon resummations.

With a future e^+e^- collider like the International Linear Collider, the $t\bar{t}$ cross section could also be measured near its threshold and the mass obtained in a way similar to the determination of the W boson mass from $W\,W$ production at LEP2 [26].

References

1. Halzen F, Martin A (1984) Quarks and leptons: an introductory course in modern particle physics. Wiley, New York
2. Peskin M, Schroeder D (1995) An introduction to Quantum Field Theory. Addison-Wesley, New York
3. The LEP collaborations, ALEPH, DELPHI, L3, OPAL, and the LEP Electroweak Working Group (1994) Combined preliminary data on Z parameters from the LEP experiments and constraints on the Standard Model, CERN-PPE/94-187
4. Abe F et al (CDF collaboration) (1995) Observation of top quark production in $p\bar{p}$ collisions with the collider detector at Fermilab. Phys Rev L 74:2626
5. Aba S et al (DØ collaboration) (1995) Observation of the top quark. Phys Rev L 74:2632
6. The Tevatron Electroweak Working Group for the CDF and DØ collaborations (2008) Combination of CDF and DØ results on the mass of the top quark. arXiv:hep-ex/0703034
7. Bigi I et al (1986) Production and decay properties of ultra-heavy quarks. Phys Lett B 181:157

8. Abe F et al (CDF collaboration) (2006) First direct limit on the top quark lifetime, CDF note 8104
9. AbazovVM, et al (DØ collaboration) (2007) Experimental discrimination between charge 2e/3 top quark and charge 4e/3 exotic quark production scenarios. Phys Rev L 98:41801
10. Kane G (2003) The dawn of physics beyond the Standard Model. Sci Am 6:71
11. LaiHL et al (CTEQ collaboration) (2000) Global QCD analysis of parton structure of the nucleon: CTEQ5 parton distributions. EurPhys J C 12:375
12. Fiedler F (2007) Precision measurements of the top quark mass. Habilitationsschrift, Ludwig-Maximilians-UniversitätMünchen
13. Kidionakis N et al (2003) Next-to-next-to-leading order soft-gluon corrections in top quark hadroproduction. Phys Rev D 68:114014
14. Kidionakis N et al (2006) Single top quark production at the FermilabTevatron: threshold resummation and finite-order soft gluon corrections. Phys Rev D 74:114012
15. Amsler C et al (Particle Data Group) (2008) The review of particle physics. Phys Lett B 667:1
16. The DØ collaboration. http://www-d0.fnal.gov/Run2Physics/top/top_public_web_pages/top_feynman_diagrams.html
17. Newton I (1687) The Principia: mathematical principles of natural philosophy. University of California Press, Berkeley
18. Collins J (1984) Renormalization: an introduction to renormalization, the renormalization group and the operator product expansion. Cambridge University Press, Cambridge
19. Smith MC (1997) Top quark pole mass. Phys Rev L 79:3825
20. The LEP Electroweak Working Group. http://lepewwg.web.cern.ch/LEPEWWG/plots/summer2008/
21. The DØ collaboration. http://www-d0.fnal.gov/Run2Physics/top/top_public_web_pages/top_public.html
22. The TEVNPH Working Group (2008) Combined CDF and DØ upper limits on Standard Model Higgs boson production at high mass (155–200 GeV) with 3.0 fb^{-1} of data, DØ note 5754
23. AbazovVM et al (DØ collaboration) (2008) Top quark mass extraction from $t\bar{t}$ cross-section measurements, DØ note 5742
24. Moch S, Uwer P (2008) Theoretical status and prospects for top quark pair production at hadron colliders, HEP-PH 0804.1476
25. Cacciari M et al (2008) Updated predictions for the total production cross sections of top and of heavier quark pairs at the Tevatron and at the LHC, HEP-PH 0804.2800
26. Miquel R (1997) WW physics at LEP2. Int J Mod Phys A12:5359

Chapter 5
The Matrix Element Method

The Matrix Element method is based on the likelihood to observe a given event under the assumption of a certain quantity to be measured, e.g. the mass of the top quark. It makes maximal use of kinematic information and yields the most precise measurement of the top mass.

To illustrate the concept of the Matrix Element method, the chapter starts with a short description of the likelihood function used. Afterwards, the parameterization of the detector response is discussed in Sect. 5.3. In the present analysis, the method has been extended to allow for taking the transverse momentum of the top pair system into account. This is described in Sect. 5.4. Sections 5.5–5.8 give technical details on the computation of the signal and background probabilities. The chapter ends with a discussion of the likelihood evaluation in Sect. 5.9 and the description of the so-called ensemble-testing procedure in Sect. 5.10.

5.1 Event Likelihood

For each selected event, the likelihood L_{evt} to observe the event is calculated as a function of the quantities α_i to be measured. Thus, maximizing the likelihood with respect to α_i yields a measurement of the parameters α_i and the method represents a general approach to extract information about any set of parameters in the event likelihood. However, in this chapter, we will restrict to the measurement of the top quark mass, so $\alpha_1 = m_{top}$. A possible extension of the method is discussed in Chap. 7.

Since it is assumed that the physics processes that lead to an observed event do not interfere, the event likelihood can be written as

$$L_{evt} = \sum_{prc} f_{prc} L_{prc}, \qquad (5.1)$$

A. Grohsjean, *Measurement of the Top Quark Mass in the Dilepton Final State Using the Matrix Element Method*, Springer Theses,
DOI: 10.1007/978-3-642-14070-9_5, © Springer-Verlag Berlin Heidelberg 2010

where L_{prc} is the likelihood for a given event to originate from a process (prc) and f_{prc} denotes the fraction of events from this process in the data sample. For the measurement presented here, it turned out to be sufficient to restrict the argument to the $t\bar{t}$ signal and the main background process $(Z \rightarrow \tau\tau) \, jj^1$, where the Z boson decays into two τ leptons. Any bias introduced by this and additional simplifying assumptions are corrected for by calibrating the final result using fully simulated Monte Carlo events, see Sect. 6.3. Uncertainties in the simulated data are then taken into account by the systematic uncertainties.

Thus, Eq. 5.1 can then be written as

$$L_{\text{evt}}(x_i; m_{\text{top}}; f_{t\bar{t}}) = f_{t\bar{t}} L_{t\bar{t}}(x_i; m_{\text{top}}) + (1 - f_{t\bar{t}}) L_{(Z \rightarrow \tau\tau)jj}(x_i), \qquad (5.2)$$

where each likelihood depends on the kinematics of the measured event x_i, while only the signal likelihood depends on the top quark mass.

In order to extract the top quark mass from a set of n events with measurements $x_1, ..., x_n$, a combined likelihood function is built from the event likelihoods,

$$L(x_1, ..., x_n; m_{\text{top}}) = \prod_{i=1}^{n} L_{\text{evt}}(x_i; m_{\text{top}}; f_{t\bar{t}}) \qquad (5.3)$$

and evaluated for different hypotheses on m_{top}. The top quark mass is finally determined by minimizing

$$- \ln L(x_1, , x_n; m_{\text{top}}; f_{t\bar{t}}) = - \sum_{i=1}^{n} \ln(L_{\text{evt}}(x_i; m_{\text{top}}; f_{t\bar{t}})) \qquad (5.4)$$

w.r.t. m_{top}.

5.2 Process Likelihood

The probability for a final state $y = (p_1, ..., p_n)$ with n partons of four-momenta p_i, $i = 1, ..., n$ to originate in a hard-scattering process of two initial-state particles is proportional to the differential cross-section $d\sigma_{\text{prc}}$, given by

$$d\sigma_{\text{prc}}(y) = \frac{(2\pi)^4 |\mathcal{M}_{\text{prc}}(y)|^2}{\varepsilon_1 \varepsilon_2 s} \, d\Phi_n. \qquad (5.5)$$

Here $\mathcal{M}_{\text{prc}}$ denotes the matrix element of the specific process, s the center-of-mass energy squared of the collider, and $\varepsilon_1, \varepsilon_2$ the momentum fraction of the colliding partons within respectively, the proton and the antiproton, compare Fig. 4.2.

---

[1] All possible combinations of particles and antiparticles are included implicitly. The notation +jets indicates that the number of jets is not determined, while jj e.g. refers to exactly two jets. $(Z \rightarrow \tau\tau) \, jj$ always includes the decay of the τ lepton into an electron or muon.

The partons are assumed to be massless, since their mass is significantly smaller than their energy. The n-body phase space $d\Phi_n$ is given by

$$d\Phi_n(\varepsilon_1\sqrt{s} + \varepsilon_2\sqrt{s}; p_1,, p_n) = \delta^4\left(\varepsilon_1\sqrt{s} + \varepsilon_2\sqrt{s} - \sum_{i=1}^{n} p_i\right)\prod_{i=1}^{n}\frac{d^3 p_i}{(2\pi)^3 2E_i}. \quad (5.6)$$

Since the initial state of the colliding partons is not known explicitly, the cross-section in Eq. 5.5 has to be convoluted with the parton density functions, f_{PDF}, and summed over all spin and flavor compositions of the colliding partons, given by

$$d\sigma_{prc}^{p\bar{p}}(y) = \int_{\varepsilon_1,\varepsilon_2,\Phi_n} \sum_{flavor,spin} d\varepsilon_1 d\varepsilon_2 f_{PDF}(\varepsilon_1) f_{PDF}(\varepsilon_2) d\sigma_{prc}(y). \quad (5.7)$$

In the experimental setup of any detector, the final-state particles are only reconstructed with a finite resolution. To take this into account, the cross-section in Eq. 5.7 is convoluted with the resolution functions $W(x, y)$, thus

$$d\sigma_{prc}^{p\bar{p}}(x) = \int d\sigma_{prc}^{p\bar{p}}(y)W(x, y). \quad (5.8)$$

The resolution functions are referred to as transfer functions and describe the probability density of a parton state y to be reconstructed as x. More details on the transfer functions can be found in Sect. 5.3. To account for the geometrical acceptance of the detector, the trigger conditions, and the selection cuts applied, the differential cross-section is normalized with the observable cross-section σ_{prc}^{obs}, given by

$$\sigma_{prc}^{obs} = \int_x dx d\sigma_{prc}^{p\bar{p}}(y)W(x, y)\Theta_{acc}(x), \quad (5.9)$$

where $\Theta_{acc}(x)$ is 1 for selected events, and 0 otherwise. Thus, the final likelihood of any generic process can be written as

$$L_{prc}(x) = \frac{1}{\sigma_{prc}^{obs}}\int_{\varepsilon_1,\varepsilon_2,\Phi_n} \sum_{flavor,spin} d\varepsilon_1 d\varepsilon_2 f_{PDF}(\varepsilon_1) f_{PDF}(\varepsilon_2) d\sigma_{prc}(y)W(x, y). \quad (5.10)$$

A schematic visualization of the signal likelihood $L_{t\bar{t}}$ is shown in Fig. 5.1. A detailed discussion of the matrix element and the calculation of the signal likelihood can be found in Sect. 5.5. The determination of the background likelihood is given in Sect. 5.7.

5.3 Description of the Detector Response

The performance of the Matrix Element method relies highly on the parameterization of the object resolution in the detector. This is accounted for using the

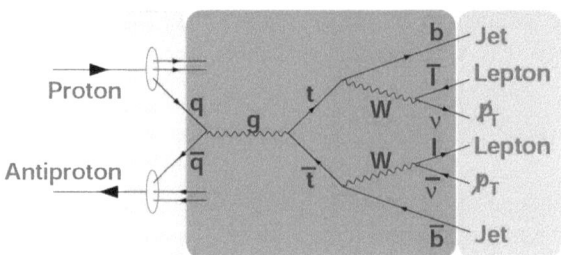

Fig. 5.1 Schematic description of the signal likelihood $L_{t\bar{t}}$. In each integration step the four-momenta of the final-state partons as well as all intermediate particles can be fully reconstructed using energy-momentum conservation, which allows the calculation of the Feynman diagram shown in *middle*. The differential cross-section is then convoluted with both the PDFs (*left*) describing the transition from the colliding particles to the initial-state partons and the transfer functions mapping the final-state partons to the measured objects (*right*)

so-called transfer functions. The transfer function $W(x, y)$ yields the likelihood to measure a partonic state y as x in the detector. Assuming that every partonic state y can be measured in the detector, the transfer function is normalized to unity

$$\int W(x, y)\mathrm{d}x = 1, \tag{5.11}$$

where the integral runs over all possible observed events x. The transfer function can be factorized into contributions from each measured final-state particle, and each factor can be handled separately.

Before the Matrix Element method can be used, two preliminary studies need to be done. First, one needs to check whether the object resolution is negligibly small, so that no transfer function is required. Secondly, if a mapping is needed, transfer functions need to be derived and their performance must be checked carefully to avoid any large offset in the calibration curve.

In order to find out which transfer function needs to be included in the Matrix Element method, the resolution of the top mass can be studied as a function of the object four-momentum to be tested. E.g., to check the resolution of the b jets, the top mass is calculated in Monte Carlo simulated PYTHIA events using the generator level information of the charged lepton and neutrino and the measured b jet momentum. Thus, any deviation of the reconstructed top mass from the generated value beyond the top width can be ascribed to the jet resolution. Figure 5.2 shows the reconstructed top mass as a function of the measured electron momentum (a), the measured muon momentum (b), and the measured b jet momentum (c) [1]. As already discussed in Sect. 2.2, the resolution of the jets is worst and yields the largest width of about 15 GeV for an assumed top mass of 175 GeV. The measurements of the charged lepton momenta are significantly better. However, the muon leads to a width of about 6 GeV, while the electron mismeasurement results in a width of only 3 GeV. The resolution of the muons becomes comparable to the one of the jets for a transverse momentum of about 80–100 GeV.

Fig. 5.2 Simulated top
quark mass distributions
for top quark decays at
$m_{\mathrm{top}} = 175$ GeV involving a
leptonic W decay, taking the
true momenta for two decay
products and the momentum
as reconstructed with the DØ
detector for the third, i.e. the
electron (**a**), the muon (**b**), or
the b jet (**c**). The mean and
width of the top quark mass
distribution are given in units
of GeV

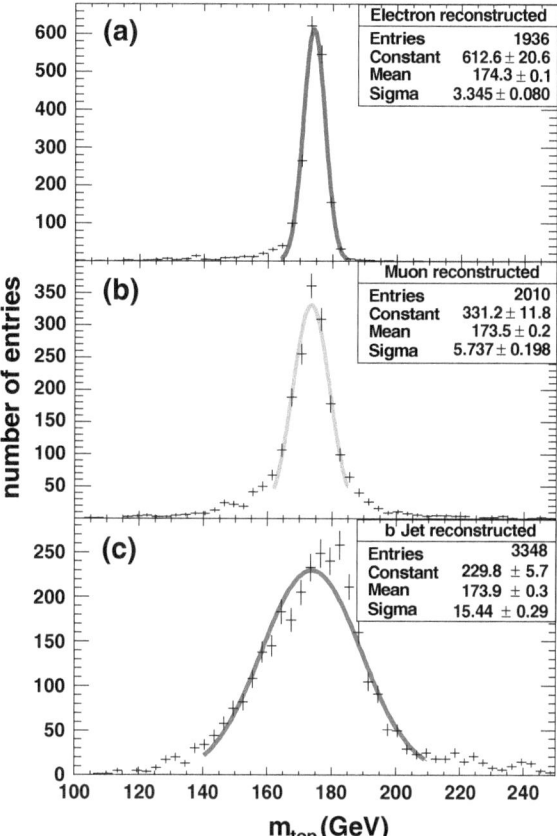

While the resolution of the electrons and jets improves slightly with the energy, as both are measured in the calorimeter and the calorimeter signal increases with the energy while noise becomes less important, the resolution of the muons degrades with the energy. Since the muon momentum is measured in the tracking system and the curvature of the track is lower for higher energies, the uncertainty increases with the muon momentum.

Besides the description of the detector resolution, transfer functions are used in the present analysis for a second purpose: the modeling of the τ lepton decay into an electron or muon in the determination of the background probabilities $(Z \to \tau\tau)\,jj$. As already discussed in Sect. 3.10, the τ decay is interfaced with TAUOLA in the generation of the background Monte Carlo events, so not part of the hard-scattering process itself. The same holds for the Zjj matrix element from the VECBOS generator which is used in the calculation of the background probabilities, see Sect. 5.7. Consequently, a separate description of the τ lepton decay is needed and an additional transfer function is applied when the background probabilities are computed.

   Finally, one has to keep in mind that each transfer function requires an additional integration in the computation of the process likelihood. Thus, one has to balance the additional computing time against the improved calibration curve.

   Hence, the following assumptions are made in the calculation of the process likelihoods.

- Electrons are assumed to be perfectly measured. Both, the direction and the energy of the detected electron are taken as the one of the partonic state.
- Since the resolution of the muon momentum becomes significantly large at high energies, only the direction of the muon is directly used in the computation of the process likelihood, while a transfer function is applied to model the resolution of the transverse muon momentum.
- As only electrons and muons emitted in the direction of flight of the τ lepton are energetic enough to pass the selection requirements, the directions of the τ lepton and the measured lepton are assumed to be the same.
- Quarks hadronize and form showers in the calorimeter, and are finally reconstructed as jets. Though it is known that the showering and the color reconnection with the beam remnant cause the direction of the jet and the quark to be different, they are treated to be the same. The effect of this assumption is small compared to the jet energy resolution, while it speeds up the computation of the process likelihoods significantly. A transfer function is applied for the jet energy. In the following, the different transfer functions used in the present analysis are discussed in-depth.

5.3.1 Parameterization of the Jet Energy Resolution

The jet energy resolution is described with a double Gaussian as a function of the energy difference ΔE between the detected jet energy E_{jet}^{det} and the assumed parton energy E_{parton}^{ass},

$$\Delta E = E_{jet}^{det} - E_{parton}^{ass}. \tag{5.12}$$

It is parameterized as

$$W_{jet}(\Delta E) = \frac{1}{\sqrt{2\pi}(p_2 + p_3 \cdot p_5)}$$
$$\times \left[\exp\left(-\frac{(\Delta E/\text{GeV} - p_1)^2}{2p_2^2} \right) + p_3 \cdot \exp\left(-\frac{(\Delta E/\text{GeV} - p_4)^2}{2p_5^2} \right) \right], \tag{5.13}$$

where the parameters p_i are themselves linear functions of the quark energy, so

$$p_i = a_i + b_i \cdot E_{parton}^{ass}/\text{GeV}. \tag{5.14}$$

Different transfer functions are derived for different eta regions $|\eta_{\text{det}}| < 0.5$, 0.5–1.0, 1.0–1.5, and 1.5–2.5 and different jet flavors, jets from light quarks (u, d, s, and c) and b quarks. In addition, b jets with a soft muon from semimuonic heavy-hadron decay are treated separately from other b jets to account for the unmeasured neutrino energy in the decay. Jets with a semielectronic decay are not considered explicitly as it is much more complicated to identify electrons within a jet.

For the sake of completeness, it has to be mentioned that Eq. 5.13 is only valid if no additional scaling factor on top of the nominal jet energy scale is varied in the fit. In case of a simultaneous measurement of the top quark mass and an additional scaling factor, this scaling factor s_{JES} has to be taken into account

$$W_{\text{jet}}(\varDelta E, s_{\text{JES}}) = \frac{W_{\text{jet}}\left(\varDelta E = \frac{E_{\text{jet}}^{\text{det}}}{s_{\text{JES}}} - E_{\text{parton}}^{\text{ass}}\right)}{s_{\text{JES}}}, \tag{5.15}$$

where the factor s_{JES} in the denominator ensures the correct normalization [2].

Recent studies by Wang [3] showed that a correction of the jet energy back to the quark energy leads to a small bias in the reconstructed W boson and top quark mass. Taking the direction of the parton-matched jets together with the energies and masses of the quarks, the mean of the reconstructed W mass is shifted by about 1.2 GeV, shown in Fig. 5.3 left. This bias can be traced back to the parton showering in the Monte Carlo due to final-state radiation. If the b quark radiates a gluon, the direction and the mass of the resulting parton shower may be significantly different from those of the b quark. Hence, using the direction of the parton-matched jets together with the mass and the energy of the parton shower, the reconstructed W mass is in excellent agreement with the generated mass shown in Fig. 5.3, right. Thus, unlike former measurements of the top quark mass with the

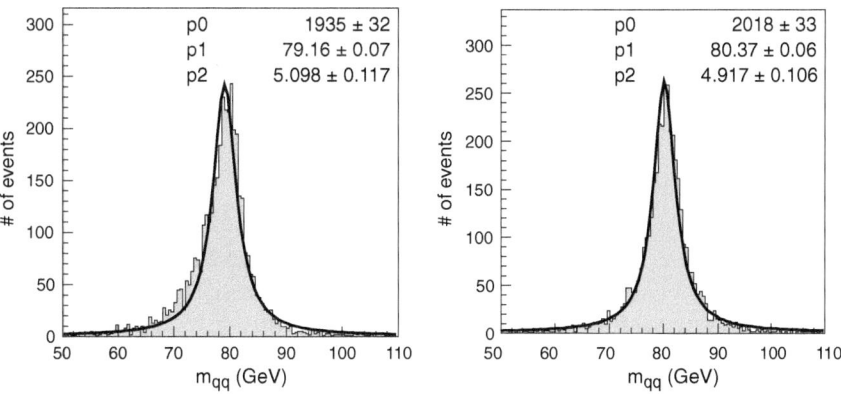

Fig. 5.3 Invariant dijet mass [3]. *Left* the quark energies and masses are used together with the jet directions to compute the W mass. *Right* the energy of the parton shower and the reconstructed jet masses are used together with the jet directions

Matrix Element method in the semileptonic decay channel, the transfer functions used in the present analysis are derived such that they correct the jet back to the level of the parton shower.

Since the two mass measurements presented in this analysis use data of two different run periods, Run IIa and Run IIb, two different sets of jet transfer functions are applied. This accounts for possible differences in the jet resolution after the detector upgrade in spring 2006. Moreover, recent studies [4] showed that the additional shifting of the jet energy derived from $Z +$ jet events, see Sect. 3.7.3, over-corrects jets from b quarks. Since jets in $Z +$ jet events dominantly arise from gluons which have a lower response and a wider shower, the shifting is no longer applied and the difference is accounted for in the systematic uncertainties. Hence, two new sets of jet transfer functions needed to be derived for the present analysis due to the improved jet handling.

The transfer function parameters are determined using fully simulated $t\bar{t}$ PYTHIA Monte Carlo events. To allow for both light and b jets in the final state, the semileptonic top decay channel is chosen. Since the event topology and the angular separation depend on the top quark mass, the transfer functions should be derived as a function of the top mass. However, due to the lack of Monte Carlo statistics, this is not feasible. Thus, samples of different generated top masses are used and the measurement has to account for the neglected top mass dependency in the calibration. For Run IIa, samples with a generated top mass of 140, 155, 160, 165, 170, 175, 180, 185, 190, 195, 200, and 215 GeV are used, for Run IIb, samples of 125, 140, 155, 160, 170, 172.5, 175, 180, 185, 190, 195, 200, 215, and 230 GeV. Besides the standard quality cuts applied in the preselection of the semileptonic decay channel [5], events are required to have exactly two jets that match the light quarks from the W boson decay, and two b jets that match the two b quarks from the top decay. For the matching, a distance ΔR of <0.5 is asked for. In addition, all jet corrections described in Sects. 3.7.1 and Sects. 3.7.3 are applied, while the cut on the jet transverse momentum is dropped.

The ten parameters to describe the energy mapping between reconstructed jets and parton shower according to Eqs. 5.13 and 5.14 are derived for each jet type in each $|\eta_{\text{det}}|$ region by minimizing the function

$$- \ln L = - \sum_{i=1}^{n} \ln W_{\text{jet}}(E_{\text{jet}}^{\text{det}}, E_{\text{parton}}^{\text{ass}}) \qquad (5.16)$$

for n selected jets. The minimization is carried out with the Minuit [6] algorithm. The performance of the algorithm improves drastically if each parameter is initialized with a sensible seed. The parameter a_3 of Eqs. 5.13 and 5.14 is fixed to 0 to constrain the fit and make sure that the two Gaussians are not inter-changeable. An additional cut of $\Delta E < 100$ GeV is applied to exclude outliers from the fit.

The full sets of parameters for the Run IIa and Run IIb transfer functions are given in Tables 5.1 and 5.2. Example plots of the b jet transfer functions which are

Table 5.1 Parameters of the jet transfer functions as defined in (5.13) and (5.14) for the Run IIa measurement for all four detector regions and the three different type of jets: light, b and μ-tagged b jets

	Light jets		b Jets		μ-Tagged b jets	
	a_i	b_i	a_i	b_i	a_i	b_i
$|\eta_{\mathrm{jet}}^{\mathrm{det}}|$: 0.0–0.5						
p_1	$-1.93 \times 10^{+0}$	1.89×10^{-3}	$-6.67 \times 10^{+0}$	7.52×10^{-3}	$6.84 \times 10^{+0}$	-1.54×10^{-1}
p_2	$4.17 \times 10^{+0}$	1.07×10^{-1}	$5.34 \times 10^{+0}$	8.09×10^{-2}	$2.48 \times 10^{+0}$	1.57×10^{-1}
p_3	$0.00 \times 10^{+0}$	2.16×10^{-4}	$0.00 \times 10^{+0}$	1.68×10^{-3}	$0.00 \times 10^{+0}$	1.32×10^{-4}
p_4	$3.34 \times 10^{+1}$	-3.35×10^{-1}	$3.92 \times 10^{+0}$	-2.11×10^{-1}	$8.38 \times 10^{+1}$	-7.73×10^{-1}
p_5	$2.61 \times 10^{+1}$	1.24×10^{-1}	$1.32 \times 10^{+1}$	1.30×10^{-1}	$2.80 \times 10^{+1}$	1.44×10^{-1}
$|\eta_{\mathrm{jet}}^{\mathrm{det}}|$: 0.5–1.0						
p_1	-2.50×10^{-1}	-1.86×10^{-2}	$-3.92 \times 10^{+0}$	-5.36×10^{-2}	$9.06 \times 10^{+0}$	-1.68×10^{-1}
p_2	$4.07 \times 10^{+0}$	1.31×10^{-1}	$4.37 \times 10^{+0}$	1.35×10^{-1}	$2.75 \times 10^{+0}$	1.64×10^{-1}
p_3	$0.00 \times 10^{+0}$	1.94×10^{-3}	$0.00 \times 10^{+0}$	2.66×10^{-4}	$0.00 \times 10^{+0}$	1.09×10^{-4}
p_4	$3.71 \times 10^{+1}$	-2.59×10^{-1}	$4.07 \times 10^{+1}$	-5.75×10^{-1}	$8.12 \times 10^{+1}$	-6.99×10^{-1}
p_5	$2.68 \times 10^{+1}$	1.49×10^{-1}	$2.93 \times 10^{+1}$	9.04×10^{-2}	$3.01 \times 10^{+1}$	1.41×10^{-1}
$|\eta_{\mathrm{jet}}^{\mathrm{det}}|$: 1.0–1.5						
p_1	$9.46 \times 10^{+0}$	-1.64×10^{-1}	4.63×10^{-1}	-1.17×10^{-1}	$1.06 \times 10^{+1}$	-1.69×10^{-1}
p_2	$3.83 \times 10^{+0}$	1.66×10^{-1}	$2.73 \times 10^{+0}$	1.87×10^{-1}	$5.59 \times 10^{+0}$	1.65×10^{-1}
p_3	$0.00 \times 10^{+0}$	3.01×10^{-3}	$0.00 \times 10^{+0}$	1.62×10^{-4}	$0.00 \times 10^{+0}$	$7.39e{-}05$
p_4	$7.93 \times 10^{+0}$	5.05×10^{-3}	$3.65 \times 10^{+1}$	-1.26×10^{-1}	$4.40 \times 10^{+1}$	$1.01 \times 10^{+0}$
p_5	$1.71 \times 10^{+1}$	1.05×10^{-1}	$3.93 \times 10^{+1}$	1.11×10^{-2}	$2.44 \times 10^{+1}$	-3.94×10^{-2}
$|\eta_{\mathrm{jet}}^{\mathrm{det}}|$: 1.5–2.5						
p_1	$1.27 \times 10^{+1}$	-1.39×10^{-1}	$7.62 \times 10^{+0}$	-1.63×10^{-1}	$1.16 \times 10^{+1}$	-1.63×10^{-1}
p_2	$4.01 \times 10^{+0}$	1.73×10^{-1}	$2.54 \times 10^{+0}$	1.89×10^{-1}	$6.69 \times 10^{+0}$	1.54×10^{-1}
p_3	$0.00 \times 10^{+0}$	3.22×10^{-4}	$0.00 \times 10^{+0}$	1.88×10^{-4}	$0.00 \times 10^{+0}$	1.09×10^{-4}
p_4	$4.66 \times 10^{+1}$	-7.82×10^{-2}	$4.40 \times 10^{+1}$	-1.44×10^{-1}	$1.28 \times 10^{+2}$	-1.81×10^{-1}
p_5	$4.69 \times 10^{+1}$	-5.29×10^{-2}	$3.81 \times 10^{+1}$	5.85×10^{-2}	$1.37 \times 10^{+2}$	-7.41×10^{-1}

Table 5.2 Parameters of the jet transfer functions as defined in (5.13) and (5.14) for the Run IIb measurement for all four detector regions and the three different type of jets: light, b and μ-tagged b jets

	Light jets		b Jets		μ-Tagged b jets	
	a_i	b_i	a_i	b_i	a_i	b_i
$|\eta_{\mathrm{jet}}^{\mathrm{det}}|$: 0.0–0.5						
p_1	$-1.47 \times 10^{+0}$	-4.91×10^{-4}	$5.15 \times 10^{+0}$	-2.32×10^{-1}	$6.79 \times 10^{+0}$	-1.64×10^{-1}
p_2	$5.10 \times 10^{+0}$	7.00×10^{-2}	$3.81 \times 10^{+0}$	1.83×10^{-1}	$2.61 \times 10^{+0}$	1.53×10^{-1}
p_3	$0.00 \times 10^{+0}$	4.36×10^{-4}	$0.00 \times 10^{+0}$	2.99×10^{-2}	$0.00 \times 10^{+0}$	2.52×10^{-4}
p_4	$2.09 \times 10^{+1}$	-2.71×10^{-1}	$-6.31 \times 10^{+0}$	2.03×10^{-2}	$3.25 \times 10^{+1}$	-3.98×10^{-1}
p_5	$1.78 \times 10^{+1}$	1.83×10^{-1}	$4.74 \times 10^{+0}$	5.00×10^{-2}	$1.89 \times 10^{+1}$	1.26×10^{-1}
$|\eta_{\mathrm{jet}}^{\mathrm{det}}|$: 0.5–1.0						
p_1	9.743×10^{-1}	-2.41×10^{-2}	$-5.517 \times 10^{+0}$	2.22×10^{-3}	$8.470 \times 10^{+0}$	-1.73×10^{-1}
p_2	$4.98 \times 10^{+0}$	9.43×10^{-2}	$6.25 \times 10^{+0}$	6.06×10^{-2}	$2.76 \times 10^{+0}$	1.61×10^{-1}
p_3	$0.00 \times 10^{+0}$	4.25×10^{-4}	$0.00 \times 10^{+0}$	2.73×10^{-3}	$0.00 \times 10^{+0}$	8.61×10^{-5}
p_4	$2.64 \times 10^{+1}$	-2.91×10^{-1}	$1.05 \times 10^{+0}$	-1.72×10^{-1}	$6.77 \times 10^{+1}$	-2.53×10^{-1}
p_5	$1.75 \times 10^{+1}$	1.82×10^{-1}	$1.40 \times 10^{+1}$	1.03×10^{-1}	$1.91 \times 10^{+1}$	4.64×10^{-2}
$|\eta_{\mathrm{jet}}^{\mathrm{det}}|$: 1.0–1.5						
p_1	$1.07 \times 10^{+1}$	-1.50×10^{-1}	$1.18 \times 10^{+1}$	-3.09×10^{-1}	$1.20 \times 10^{+1}$	-1.82×10^{-1}
p_2	$3.17 \times 10^{+0}$	1.50×10^{-1}	$2.13 \times 10^{+0}$	1.49×10^{-1}	$3.86 \times 10^{+0}$	1.56×10^{-1}
p_3	$0.00 \times 10^{+0}$	1.97×10^{-3}	$0.00 \times 10^{+0}$	1.09×10^{-2}	$0.00 \times 10^{+0}$	2.61×10^{-4}
p_4	$1.47 \times 10^{+1}$	-1.53×10^{-3}	$2.70 \times 10^{+0}$	-4.14×10^{-2}	$3.57 \times 10^{+1}$	2.67×10^{-1}
p_5	$1.73 \times 10^{+1}$	5.02×10^{-2}	$1.14 \times 10^{+1}$	7.44×10^{-2}	$2.04 \times 10^{+2}$	$-1.37 \times 10^{+0}$
$|\eta_{\mathrm{jet}}^{\mathrm{det}}|$: 1.5–2.5						
p_1	$2.08 \times 10^{+1}$	-2.25×10^{-1}	$1.66 \times 10^{+1}$	-2.65×10^{-1}	$1.77 \times 10^{+1}$	-2.09×10^{-1}
p_2	$3.79 \times 10^{+0}$	1.35×10^{-1}	$3.87 \times 10^{+0}$	1.30×10^{-1}	$6.57 \times 10^{+0}$	1.36×10^{-1}
p_3	$0.00 \times 10^{+0}$	2.77×10^{-3}	$0.00 \times 10^{+0}$	3.91×10^{-3}	$0.00 \times 10^{+0}$	1.29×10^{-4}
p_4	$2.48 \times 10^{+1}$	-5.16×10^{-2}	$1.28 \times 10^{+1}$	-4.41×10^{-2}	$1.01 \times 10^{+2}$	-2.83×10^{-1}
p_5	$1.96 \times 10^{+1}$	4.18×10^{-2}	$1.84 \times 10^{+1}$	5.34×10^{-2}	-1.72×10^{-2}	1.24×10^{-1}

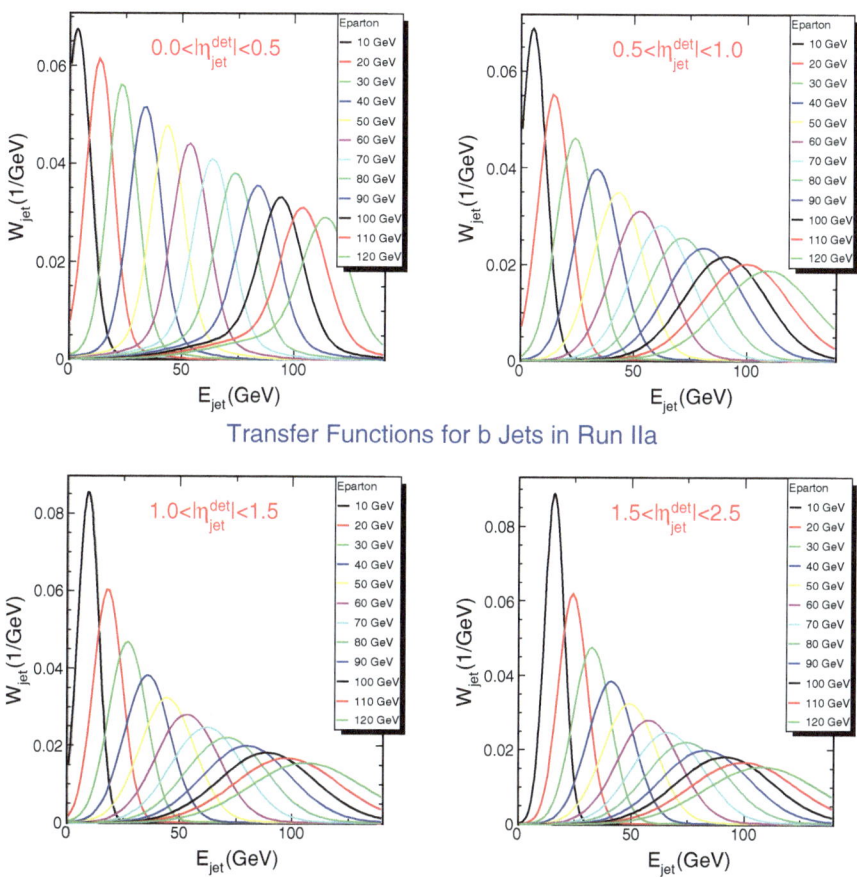

Fig. 5.4 Jet energy transfer functions as defined in (5.13) for b jets in the four different detector regions for parton energies of 10–120 GeV in steps of 10 GeV for Run IIa

dominantly used in the computation of the signal likelihood are shown for different quark energies in Figs. 5.4 for Run IIa, and 5.5 for Run IIb.

To check the performance of the transfer functions, two important cross checks are carried out. First, the energy difference between the jets and the partons is compared to the energy difference from the partons and the smeared partons using the transfer functions. If the transfer functions describe the energy resolution in the Monte Carlo simulated events, both distributions should agree. An example for this kind of comparison is shown for central b jets in Fig. 5.6; all comparison plots can be found in Ref. [7]. The blue histogram corresponds to the ΔE distribution from Monte Carlo simulated events, the red histogram, to the one generated by the transfer functions. To check the performance in the different energy bins, the comparison plots are in bins of the parton energy: less than 40, 40–60, 60–80,

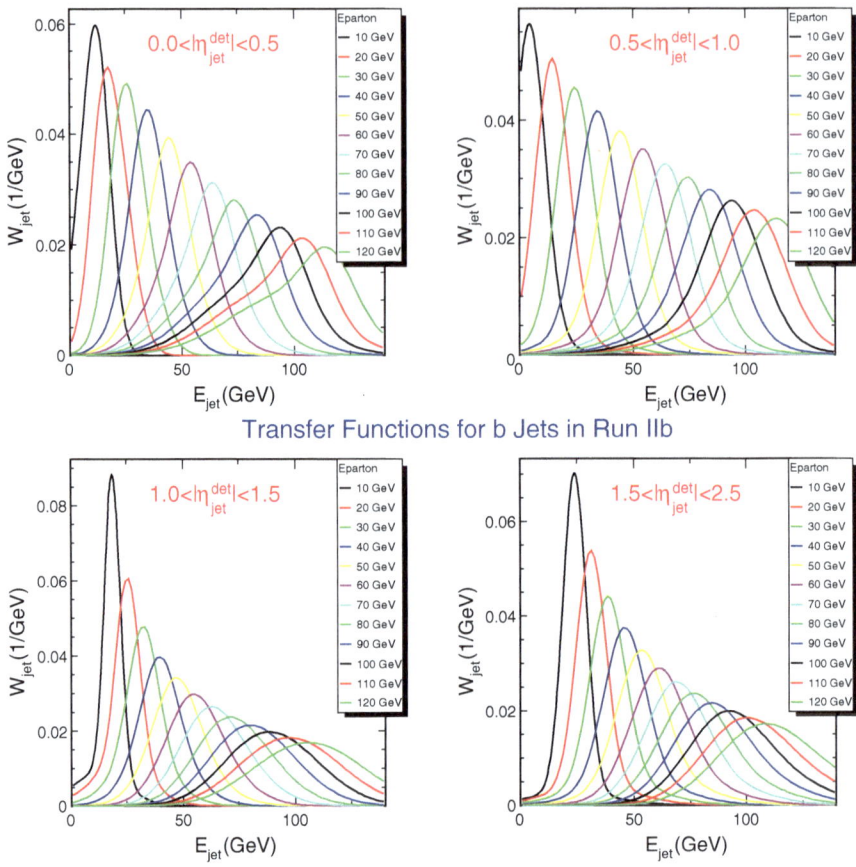

Fig. 5.5 Jet energy transfer functions as defined in (5.13) for b jets in the four different detector regions for parton energies of 10–120 GeV in steps of 10 GeV for Run IIb

80–100, 100–120, and more than 120 GeV. In a second test, the W boson and the top quark mass are calculated using the measured jet energies and the smeared parton energies together with the partonic lepton and neutrino four-momenta. A certain offset is added to the masses from the smeared parton energies and a likelihood is calculated that compares the shifted mass to the reference histogram. The χ^2 of this comparison is fitted with a parabola to determine the most likely offset and its uncertainty. The W boson and top quark masses measured in this way are plotted as a function of the generated top mass to check the performance in the different Monte Carlo samples. The results for the Run IIa and Run IIb jet transfer functions are shown in Fig. 5.7.

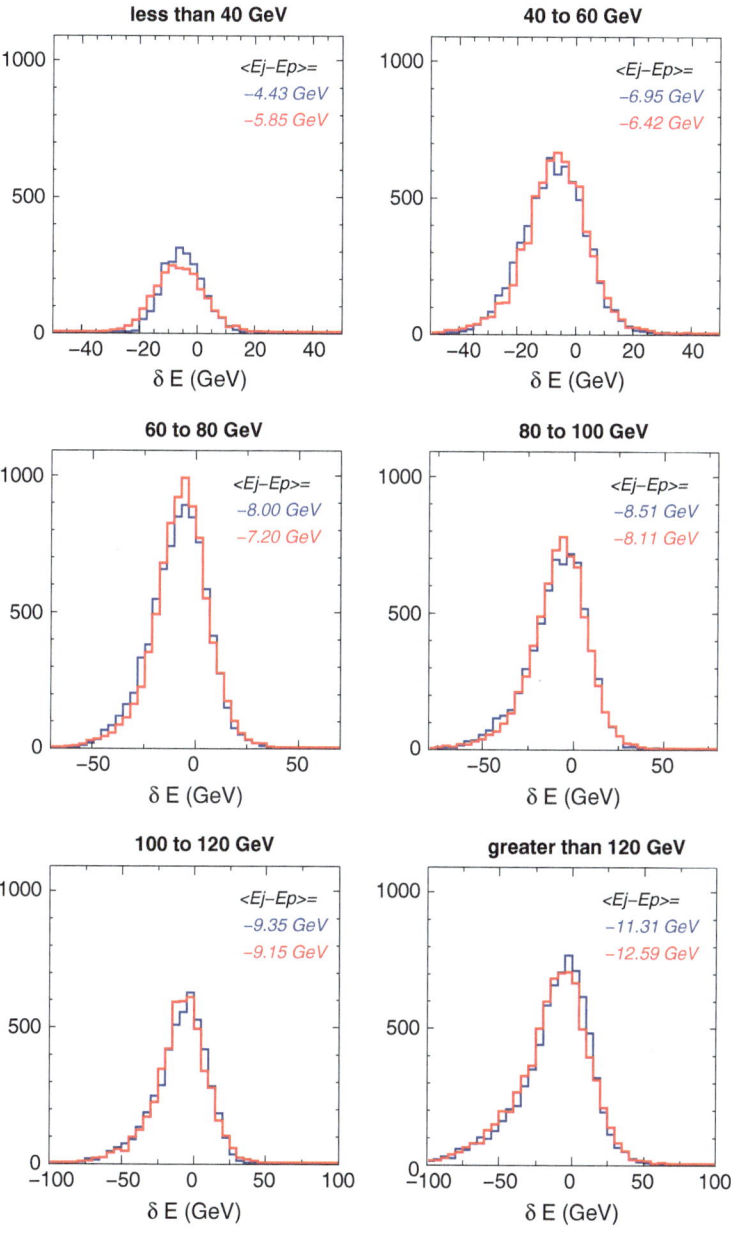

Fig. 5.6 $\delta E = E_{\text{jet}}^{\text{det}} - E_{\text{parton}}^{\text{ass}}$ distribution for b jets with $|\eta| < 0.5$. The *blue histogram* corresponds to the δE distribution from Monte Carlo, the *red histogram* to the one generated by the transfer functions

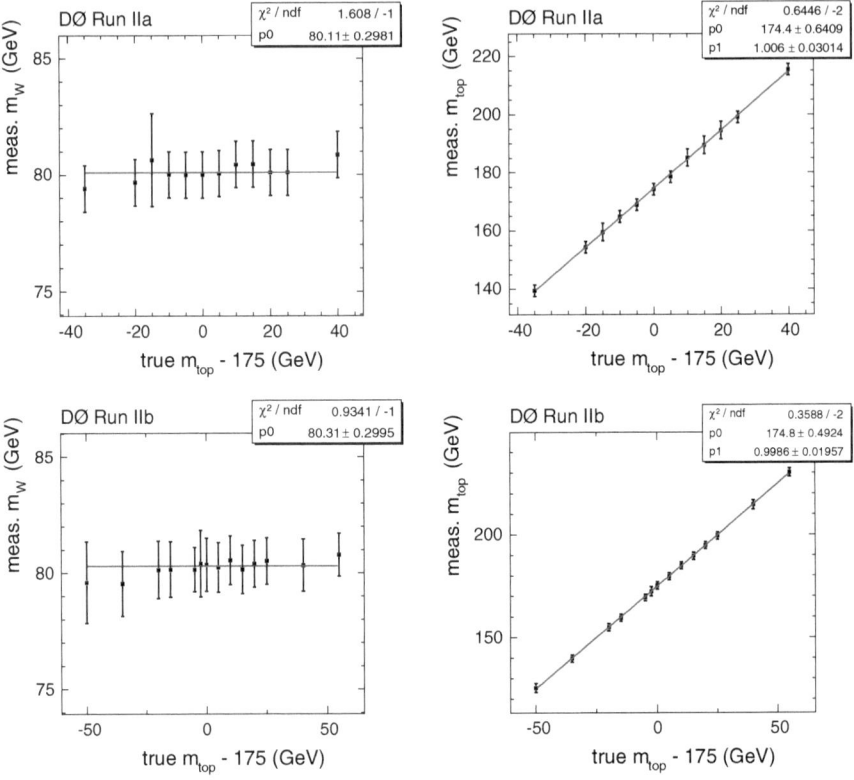

Fig. 5.7 Measured W boson and top quark masses as a function of the generated top mass to check the performance of the jet transfer functions; *top* for Run IIa; *bottom* for Run IIb

5.3.2 Parameterization of the Muon Momentum Resolution

The resolution of a muon in the central tracking chamber is described with a single Gaussian as a function of the difference

$$\Delta\left(\frac{q}{p_T}\right) = \left(\frac{q}{p_T}\right)^{det}_{\mu} - \left(\frac{q}{p_T}\right)^{ass}_{\mu} \tag{5.17}$$

between the muon charge q divided by its transverse momentum p_T for the assumed (ass) and detected (det) muon. This parameterization reflects the measurement of the curvature in the tracking detector which is proportional to q/p_T. The muon transfer function is given by

$$W_{\mu}\left(\Delta\left(\frac{q}{p_T}\right)\right) = \frac{1}{\sqrt{2\pi}\sigma}\exp\left(-\frac{1}{2}\left(\frac{(q/p_T)^{det}_{\mu} - (q/p_T)^{ass}_{\mu}}{\sigma}\right)^2\right), \tag{5.18}$$

Fig. 5.8 Example for the dependency of the muon resolution on the pseudorapidity η_μ^{det} and the inverse transverse momentum $(1/p_T)_\mu^{\text{det}}$ [1]

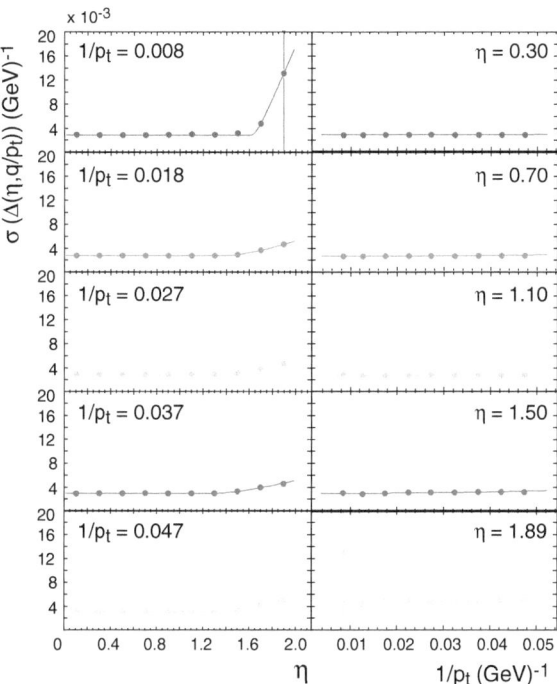

where the resolution σ mostly depends on the measured pseudorapidity for non-central muons with a pseudorapidity $\eta_\mu^{\text{det}} > 1.4$, while the dependency on the measured transverse momentum $(1/p_T)_\mu^{\text{det}}$ is only weak, see Fig. 5.8.

The dependency on the pseudorapidity can be explained by the geometry of the tracking detector. In the forward region, tracks can no longer be measured in all layers, and the resolution degrades. Thus, the resolution σ is parameterized as

$$\sigma = \begin{cases} \sigma_0 & \text{for } |\eta| \leq \eta_0 \\ \sqrt{\sigma_0^2 + [c(|\eta| - \eta_0)]^2} & \text{for } |\eta| > \eta_0 \end{cases}. \qquad (5.19)$$

To allow for a weak $(1/p_T)_\mu^{\text{det}}$ dependency, the parameters σ_0 and c are themselves linear functions of the inverse momentum, so

$$\sigma_0 = a_{\sigma_0} + b_{\sigma_0} \left(\frac{1}{p_T}\right)_\mu^{\text{det}} \qquad (5.20)$$

$$c = a_c + b_c \left(\frac{1}{p_T}\right)_\mu^{\text{det}}. \qquad (5.21)$$

Since muons are not required to create hits in both parts of the tracking system, SMT and CFT, about 1% of the muons have no inner hits in the SMT detector.

Table 5.3 Muon transfer function parameters for muons with and without SMT hits for the Run IIa pre- and post-shutdown periods, as well as for Run IIb

	$n_{\mathrm{SMT}} = 0$	$n_{\mathrm{SMT}} > 0$
Run IIa pre		
$a_{\sigma_0}(1/\mathrm{GeV})$	5.23×10^{-3}	3.16×10^{-3}
$b_{\sigma 0}$	-5.27×10^{-2}	-2.77×10^{-2}
$a_c\ (1/\mathrm{GeV})$	2.04×10^{-2}	4.24×10^{-3}
b_c	-1.73×10^{-1}	1.38×10^{-1}
η_0	$1.40 \times 10^{+0}$	$1.40 \times 10^{+0}$
Run IIa post		
$a_{\sigma_0}(1/\mathrm{GeV})$	4.76×10^{-3}	3.16×10^{-3}
$b_{\sigma 0}$	-3.11×10^{-2}	-2.77×10^{-2}
$a_c\ (1/\mathrm{GeV})$	2.07×10^{-2}	4.24×10^{-3}
b_c	-1.78×10^{-1}	1.38×10^{-1}
η_0	$1.40 \times 10^{+0}$	$1.40 \times 10^{+0}$
Run IIb		
$a_{\sigma_0}(1/\mathrm{GeV})$	3.62×10^{-3}	2.08×10^{-3}
b_{σ_0}	1.39×10^{-2}	1.13×10^{-2}
$a_c\ (1/\mathrm{GeV})$	2.07×10^{-2}	7.67×10^{-3}
b_c	7.04×10^{-2}	7.85×10^{-2}
η_0	$1.40 \times 10^{+0}$	$1.40 \times 10^{+0}$

Thus, the track curvature is less precisely known, and the resolution is worse than for muon tracks with SMT hits. Although the resolution of these muons can be improved by constraining the muon track to the primary vertex, the resolution is still different and two different sets of parameters are derived for muons with and without hits in the SMT detector.

Additionally, one has to account for the two major shutdowns of the DØ detector in fall 2004 and spring 2006. After the first shutdown, the resolution of the muons degraded due to the reduction of the magnetic field, while it improved again after the second due to the additional layer in the inner tracker, see Sect. 2.2.2. Hence, three different sets are derived for the different run periods referred to as Run IIa pre, Run IIa post and Run IIb.

Using muon tracks from simulated PYTHIA $t\bar{t}$ and Z + jets events, the parameters listed in Table 5.3 are found. More details on the determination of the muon transfer function are given by Haefner et al. [1].

5.3.3 Parameterization of the τ Lepton Decay

The τ transfer function $W_\tau(E_\ell^{\mathrm{ass}}, E_\tau^{\mathrm{ass}})$ yields the probability density for a lepton (electron or muon) to have the energy E_ℓ^{ass} given the true τ energy E_τ^{ass}. It can be parameterized as a 3rd-order polynomial in the ratio of both energies:

$$W_\tau(E_\ell^{\mathrm{ass}}, E_\tau^{\mathrm{ass}}) = \sum_{i=0}^{3} a_i \left(\frac{E_\ell^{\mathrm{ass}}}{E_\tau^{\mathrm{ass}}}\right)^i . \tag{5.22}$$

As the τ transfer functions are used to model the $Z \to \tau\tau$ decay, it is assumed that every τ lepton decays to an electron or muon and the probability density is normalized to unity, so

$$\int\limits_0^1 d\left(\frac{E_\ell^{\mathrm{ass}}}{E_\tau^{\mathrm{ass}}}\right) W_\tau(E_\ell^{\mathrm{ass}}, E_\tau^{\mathrm{ass}}) = 1. \qquad (5.23)$$

The τ transfer functions are derived using ALPGEN $Z \to \tau\tau$ Monte Carlo events. The standard corrections described in Sects. 3.3 and 3.4 are applied and a loose selection is used to retain enough statistics. The partonic $Z \to \tau\tau \to \ell\ell$ decay is reconstructed using Monte Carlo information, and the ratio of the partonic lepton energy to the partonic τ energy is determined. The result is shown in Fig. 5.9, and a polynomial of 3rd-order is used to fit the distribution. The parameters of the normalized τ transfer functions are given in Table 5.4. As the τ lepton decays into three particles, one charged lepton and two neutrinos, on average the charged lepton carries a third of the τ energy. In case the charged lepton is emitted in opposition to the direction of flight of the τ lepton, the resulting lepton energy is close to zero, while it is maximal when emitted in the same direction.

To check the stability of the τ transfer function against the direction and the transverse momentum of the τ lepton, the transfer functions are derived for different η and p_T bins. Within the statistical uncertainties, no discrepancy is observed. As the τ transfer function does not depend on detector effects, the same transfer function can be used for the Run IIa and Run IIb measurement.

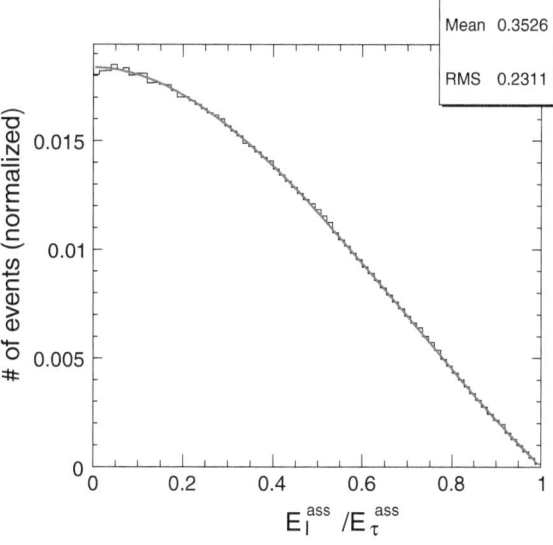

Fig. 5.9 Probability for a lepton to carry a given energy fraction of the initial τ lepton energy

Table 5.4 Parameters of the
τ transfer function

a_0	1.84×10^{-2}
a_1	-5.55×10^{-5}
a_2	-3.50×10^{-2}
a_3	1.66×10^{-2}

5.4 Parameterization of the Top Pair Transverse Momentum

A significant shift of the calibration curve is observed in the Matrix Element method when events with additional jets in the final state are included [8]. Thus, in the semileptonic analysis exactly four jets are required [9]. However, this solution is not reasonable in the dilepton channel as the size of the data sample is already low and requiring exactly two jets increases the statistical uncertainty considerably. Hence, a different approach is chosen.

It is based on the idea that additional jets in the event arise either from initial-state radiation (ISR) or final-state radiation (FSR). In case of FSR, a quark from the top quark decay emits an additional gluon, leading to two jets which share the energy of the original quark. In case of ISR, one of the colliding quarks radiates a gluon and the top quarks produced are no longer balanced in the transverse plane.

To study the effect of additional jets, ALPGEN $t\bar{t}$ Monte Carlo events are used. As the dependency of additional jets on the top mass is expected to be low, samples of different generated top masses, 150, 160, 165, 170, 175, 180, 185, and 190 GeV are merged for this study. In a first step, the top pair transverse momentum is calculated using the generated top quarks from Monte Carlo information and the distributions are plotted for events with exactly two and more than two reconstructed jets in the final state, see Fig. 5.10.

As expected, the transverse momentum of the top pair system peaks at low values in the case of exclusively two jets, left figure. With a spread of about 5 GeV, this peak is not exactly at zero.

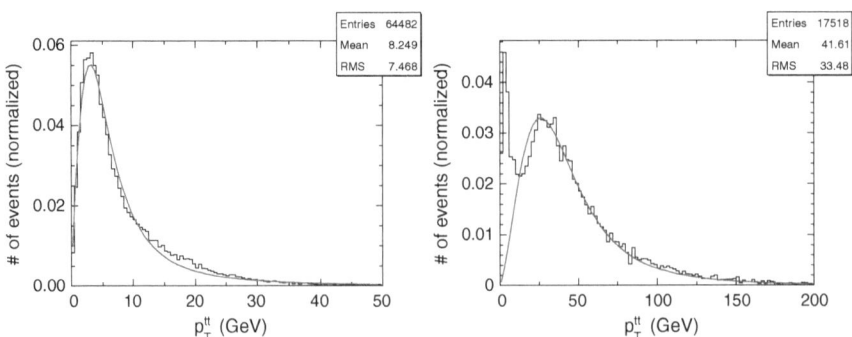

Fig. 5.10 Top pair transverse momentum in $t\bar{t}$ ALPGEN Monte Carlo events; *left* in the case of exactly two reconstructed jets; *right* in the case of more than two jets

In the case of more than two reconstructed jets, two peaks are observed, right figure. The first peak arises from final-state radiation and additional collisions. Indeed those events have an additional jet but no significant amount of transverse momentum is observed. To handle them properly, all possible pairs of jets from quark and corresponding gluon have to be merged and treated as one in the integration. Since the method would become significantly slower and the number of events with FSR is negligibly small compared to the number of events with ISR due to the so-called dead cone effect [10], the first peak is ignored in the fit and only the two leading jets are considered when calculating the signal likelihood. The second peak in the right figure arises from events with initial-state radiation that leads to a significant amount of transverse momentum. Thus, the additional jet from initial-state radiation can be simulated by integrating over the top pair transverse momentum.

To validate these assumptions, the top quark masses for events with exactly three jets and a generated top mass of 170 GeV are calculated. The top mass is reconstructed by combining the W boson from Monte Carlo truth information and, respectively, one or two measured jets. The two reconstructed jets are required to match the partonic b jets and the third jet is added to the closest b jet. The blue curve in Fig. 5.11 shows the result for events in the first peak of the top pair transverse momentum distribution, the red, in the second peak. The fact that the offset in events from the second peak is about 50 GeV, while it is only 15 GeV in events from the first peak, makes clear that the third jet in these events really comes from FSR.

Consequently, the left distribution in Fig. 5.10 and the second peak in the right distribution are fitted using the following functional form

$$W_{p_T^{t\bar{t}}} = a_0 \cdot \frac{\left(p_T^{t\bar{t}}\right)^{a_1}}{\left(p_T^{t\bar{t}} \cdot p_T^{t\bar{t}} + a_2\right)^{a_3}}, \tag{5.24}$$

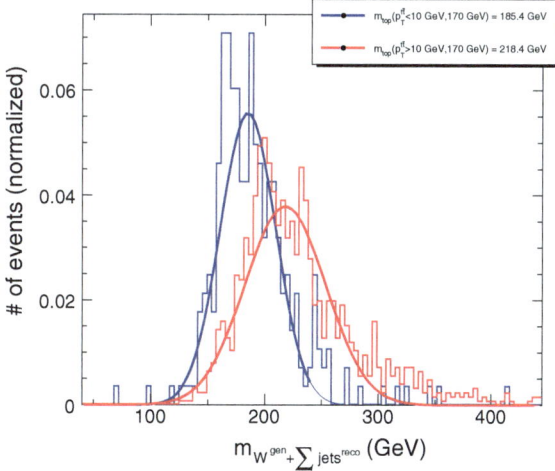

Fig. 5.11 Reconstructed top quark masses in events with more than two jets from the first (*blue*) and the second (*red*) peak of the transverse top pair momentum distribution, see Fig. 5.10. The top quark masses are calculated by combining the W boson from Monte Carlo truth information and, respectively, one or two measured jets. The two reconstructed jets are required to match the partonic b jets. The third jet is added to the closest b jet

Table 5.5 Fitted parameters	$n_{jet} = 2$	$n_{jet} > 2$
for the normalized $p_T^{t\bar{t}}$		
probability density as given		
in (5.24) a_0 (GeV)	$2.96 \times 10^{+1}$	$7.02 \times 10^{+5}$
a_1 (GeV)	8.85×10^{-1}	$1.43 \times 10^{+0}$
a_2 (GeV^2)	$2.88 \times 10^{+1}$	$1.99 \times 10^{+3}$
a_3 (GeV^2)	$1.80 \times 10^{+0}$	$2.80 \times 10^{+0}$

with parameters a_0, a_1, a_2, and a_3 to be determined. The results from these fits are normalized such that

$$\int_0^\infty d\left(p_T^{t\bar{t}}\right) W_{p_T^{t\bar{t}}} = 1. \tag{5.25}$$

The final parameters for the normalized $p_T^{t\bar{t}}$ probability are listed in Table 5.5. It has been checked that the parameterization does not depend on the event selection and is valid for both Run IIa and Run IIb.

It has to be noticed that the transfer functions, discussed in Sect. 5.3, are conceptually different from the probability density of the top pair transverse momentum, $p_T^{t\bar{t}}$. The transfer functions describe the compatibility of a measured quantity to an assumed quantity in a given integration step. The top pair transverse momentum, however, is not related to any quantity measured in the detector, like the unclustered energy defined in Sect. 3.8 as this quantity has a large systematic uncertainty [11].

5.5 Calculation of the Signal Likelihood

In analogy to the likelihood of any generic process described in Eq. 5.10, the signal likelihood $L_{t\bar{t}}$ is given by

$$L_{t\bar{t}}(x; m_{top}) = \frac{1}{\sigma_{t\bar{t}}^{obs}(m_{top})} \int_{\varepsilon_1, \varepsilon_2, \Phi_6, p_T^{t\bar{t}}} \sum_{flavor,spin} d\varepsilon_1 d\varepsilon_2 f_{PDF}(\varepsilon_1) f_{PDF}(\varepsilon_2)$$

$$\times \frac{(2\pi)^4 |\mathcal{M}_{t\bar{t}}(y; m_{top})|^2}{\varepsilon_1 \varepsilon_2 s} d\Phi_6 W(x, y) dp_T^{t\bar{t}} W_{p_T^{t\bar{t}}}, \tag{5.26}$$

where $W(x, y)$ summarizes the resolution functions of the jets (5.13) and the muons (5.18) and where the probability density for the transverse top pair momentum $W_{p_T^{t\bar{t}}}$ is included. Even though top quark pairs can be produced via both quark–antiquark annihilation and gluon–gluon fusion, the latter production mechanism is not taken into account here. This is justified because at the Tevatron only 15% of the top quark pairs are produced in gluon–gluon fusion. Additionally, the top quark and the W boson propagators, as well as the decay parts of the matrix element which contain most of the information on the top mass are identical. The effect of the spin correlation between the top quarks is not considered in the matrix element, as it does not affect the mass systematically. Thus, the leading-order matrix element [12] for the signal process $q\bar{q} \to t\bar{t} \to y$ is given by

$$|\mathcal{M}_{q\bar{q} \to t\bar{t} \to y}|^2 = \frac{g_s^4}{9} F\bar{F} \left(2 - \beta^2 s_{qt}^2\right), \tag{5.27}$$

where $g_s^2/4\pi$ is the strong coupling constant, β, the velocity of the top quarks in the top pair rest frame, and s_{qt}, the sine of the angle between the incoming parton and the outgoing top quark in the top pair rest frame. The factors F and $\bar{F}$ describe the kinematics of the top and antitop quark decay. In the case of two leptonically decaying W bosons, they are given by Ref. [12]

$$F = \frac{g_w^4}{4} \left(\frac{m_{b\ell\nu}^2 - m_{\ell\nu}^2}{\left(m_{b\ell\nu}^2 - m_{\text{top}}^2\right)^2 + \left(m_{\text{top}} \Gamma_{\text{top}}\right)^2} \right) \left(\frac{m_{b\ell\nu}^2 \left(1 - \hat{c}_{b\ell}^2\right) + m_{\ell\nu}^2 \left(1 + \hat{c}_{b\ell}\right)^2}{\left(m_{\ell\nu}^2 - m_W^2\right)^2 + (m_W \Gamma_W)^2} \right) \tag{5.28}$$

$$\bar{F} = \frac{g_w^4}{4} \left(\frac{m_{\bar{b}\ell\bar{\nu}}^2 - m_{\ell\bar{\nu}}^2}{\left(m_{\bar{b}\ell\bar{\nu}}^2 - m_{\text{top}}^2\right)^2 + \left(m_{\text{top}} \Gamma_{\text{top}}\right)^2} \right) \left(\frac{m_{\bar{b}\ell\bar{\nu}}^2 \left(1 - \hat{c}_{\bar{b}\ell}^2\right) + m_{\ell\bar{\nu}}^2 \left(1 + \hat{c}_{\bar{b}\ell}\right)^2}{\left(m_{\ell\bar{\nu}}^2 - m_W^2\right)^2 + (m_W \Gamma_W)^2} \right). \tag{5.29}$$

Here, g_w denotes the weak charge, m_{top} and m_W, the masses of the top quark and the W boson, Γ_{top} and Γ_W, their widths. The invariant masses of the top and W in a particular event are denoted by m_{xyz} and m_{yz}, respectively, where x, y, and z are the decay products. $\hat{c}_{xy}$ is the cosine of the angle between particles x and y in the W rest frame.

The width of the top quark can be written as a function of its mass [13]

$$\Gamma_{\text{top}} = \frac{G_F m_{\text{top}}^3}{8\pi \sqrt{2}} \left(1 - \frac{m_W^2}{m_{\text{top}}^2}\right)^2 \left(1 + 2\frac{m_W^2}{m_{\text{top}}^2}\right) \left[1 - \frac{2\alpha_s}{3\pi}\left(\frac{2\pi^2}{3} - \frac{5}{2}\right)\right], \tag{5.30}$$

where G_F denotes the Fermi coupling.

The computation of the signal likelihood involves an integral over all possible momenta of the colliding partons and the 6-body phase space of the final-state particles, cf. Eq. 5.26. Additionally, the correct association of the final-state jets to the partons in Eqs. 5.28 and 5.29 is not known, and the signal likelihood also needs to be computed for both jet-parton assignments using the two leading jets. To make a numerical calculation of the integral feasible, no integration is performed over quantities that are assumed to be well measured: the jet directions and the charged lepton directions. In addition, the energy of the electron is assumed to be exactly known. After these assumptions, an integration over 8 dimensions (+1 for each muon) remains, as is shown in Table 5.6.

Table 5.6 The number of dimensions in the signal probability integration for dilepton $t\bar{t}$ events

Energy fractions ε_1, ε_2 of the colliding partons:	2
x- and y-momentum of the top pair system	2
6 Final-state partons (of known mass):	18
2 Jet directions:	-4
2 Charged lepton 3-momenta:	-6
Energy and momentum conservation	-4
Remaining dimensions:	8

The following three considerations have been taken into account when computing the integral:

- To minimize the computing time needed to perform the remaining integration numerically, it is desirable to transform the integration variables such that the integrand is strongly peaked as a function of single integration variables.
- At the same time, it should be possible to easily reconstruct the event kinematics, i.e. the four-momenta of the partons in the event, from the integration variables. The gain from the transformation would be lost if this step involved solving too complicated equations. E.g. using one invariant W mass in addition to two top masses the event kinematics can no longer be solved with simple analytic equations.
- The integration variables should be uncorrelated.

Consequently, the following eight integration variables ($+1$ for each muon) have been chosen:

$$|\vec{p}_{b_1}|, |\vec{p}_{b_2}|, p_x^{\nu_1} - p_x^{\nu_2}, p_y^{\nu_1} - p_y^{\nu_2}, p_x^{t\bar{t}}, p_y^{t\bar{t}}, m_{top_1}^2, m_{top_2}^2 (, (q/p_T)_\mu). \qquad (5.31)$$

The boundaries of the integration variables are adjusted such that they cover the possible range at the Tevatron; e.g. the absolute b jet momentum cannot exceed half the center-of-mass energy. The integration over the top pair momentum $p_x^{t\bar{t}}, p_y^{t\bar{t}}$ runs from -250 to 250 GeV.

The matrix element contains a Breit-Wigner peak associated with each top quark mass which is so narrow that it is possible to perform the integration over $m_{top_1}^2$ and $m_{top_2}^2$ analytically in the narrow width approximation using the relation

$$\int \frac{f(m^2)\mathrm{d}(m^2)}{(m^2 - m_0^2)^2 + (m_0\Gamma_0)^2} \simeq \frac{\pi f(m_0^2)}{m_0\Gamma_0}, \qquad (5.32)$$

where $f(m^2)$ denotes any reasonably behaved function of the integration variable m^2, and m_0 and Γ_0 are the mass and width of the resonance, respectively. Taking this into account, the dimension of the final integral can be reduced by two.

To summarize the calculation of the signal likelihood $L_{t\bar{t}}$, the computation at each point in the integration space for the two jet-parton assignments is summarized in the following:

- The four-momenta of the final-state particles are reconstructed from the integration variables, the measured jet and lepton angles, and the electron energy, as derived in Appendix A.
- The matrix element is computed following Eqs. 5.27, 5.28, and 5.29.
- The parton distribution functions that have been applied to produce the Monte Carlo samples employed are evaluated. E.g. deriving the calibration curve in case of the full detector simulation, the CTEQ6L1 parton distribution functions are used.
- The transfer functions are applied to account for the probabilities to observe the measured jet energies and muon transverse momentum given the energies and momentum reconstructed in the first step.
- The probability to observe the assumed top pair transverse momentum is evaluated.
- In a final step, the Jacobian determinant derived in Appendix B for the transformation from momenta in Cartesian coordinates to the integration space, Eq. B.8 is included.

The remaining integration is performed on Monte Carlo basis with the numerical integration program VEGAS [14, 15]. The interface to the VEGAS integration algorithm is provided by the GNU Scientific Library (GSL) [16].

5.6 Normalization of the Signal Likelihood

To obtain the final signal likelihood given in Eq. 5.26, the observable cross-section

$$\sigma_{t\bar{t}}^{\mathrm{obs}} = \int\limits_{y,x} \mathrm{d}x \mathrm{d}\sigma_{\mathrm{prc}}^{p\bar{p}}(y) W(x,y) \Theta_{\mathrm{acc}}(x) \tag{5.33}$$

needs to be computed. Here, $\Theta_{\mathrm{acc}}(x)$ denotes the detector acceptance and the kinematic selection cuts applied. As the spectrum of the final-state particles, and thus the selection efficiency, depends on the top mass, omitting the acceptance would lead to a significant bias. Additional cuts that do not depend on the top quark mass do not need to be considered since they do not affect the shape of the normalization curve. To be more precise, the acceptance function Θ_{acc} yields

- 0 if any top mass dependent selection cut fails;
- 0 if the distance $\Delta R(j, j')$ between any pair of jets is <0.5;
- 0 if the distance $\Delta R(j, \ell)$ between any jet and charged lepton is <0.5.

The latter two cuts account for the possibility that two particles can overlap in the calculation of the integral while this is not true for reconstructed objects in the detector. If no selection cuts are applied, the observable cross section corresponds to the leading order cross section for $t\bar{t}$ production. The branching ratio of the top quark decay is not included as an overall scaling factor is compensated by the relative background-to-signal normalization, see Sect. 5.8.

It has been checked that the transverse momentum of the top pair system does not have to be accounted for in the normalization and can be assumed to vanish as it does not depend on the top quark mass. As the masses of the final-state particles are know, the two additional constraints help to reduce the dimensions of the integral from 24, the four-momenta of the 6 final-state particles, to 16. Since the kinematic selection does not apply on the partons but on the measured particles, the transfer functions are used to map between them. Thus, $\sigma_{t\bar{t}}^{\text{obs}}$ can be calculated by adding the jet energies and the inverse muon momenta as additional integration variables and interpreting the acceptance Θ_{acc} as a function of the jet energies and the inverse muon momenta. In the calculation of the observable cross section, the missing transverse energy is assumed to be identical to the sum of the neutrino momenta, so

$$\not{E}_x = -\left(\sum_{i=1}^{2} p_x^{j_i} + p_x^{\ell_i}\right) \tag{5.34}$$

$$\not{E}_y = -\left(\sum_{i=1}^{2} p_y^{j_i} + p_y^{\ell_i}\right) \tag{5.35}$$

The integral is computed using VEGAS [14, 15]. To maximize the performance of the algorithm, the top quark and W boson masses with their Breit-Wigner peaks are used as integration variables.

The final normalization curve as a function of the top quark mass is fitted with a 3rd-order polynomial.

5.7 Calculation of the Background Likelihood

Ideally, the likelihood for a given event to be observed should be calculated for every possible background process that could give rise to the event, and included in the analysis. However, it has already been shown in the first Run IIa measurement [9] in the semileptonic final state that such a complete treatment is not necessarily mandatory. In that analysis, only the $Wjjjj$ likelihood was computed, and the differences between W + jets and multijet events were neglected. The dilepton samples are even purer than the semileptonic samples, and thus similar simplifications can be expected to hold. Tests have been performed with likelihoods calculated for various background processes. The criterion to decide whether or not a certain background process needs to be taken into account in the event likelihood calculation according to Eq. 5.2 was that a calibration curve as close to optimal as possible should be obtained with as few background processes as possible to save CPU time. It turns out that only the main background channel needs to be accounted for explicitly: the production of Z bosons in association with two light quark jets[2].

Because many different Feynman graphs contribute to this process, the leading-order Zjj matrix element times the parton distribution function factors is evaluated using the VECBOS [17] Monte Carlo generator. VECBOS takes the relative importance of the various subprocesses into account and performs a statistical sampling of all possible spin, flavor, and color configurations. However, in case of an intermediate τ lepton decay, i.e. $(Z \rightarrow \tau\tau)\,jj$ events, this decay is not described by VECBOS and needs to be modeled with the τ transfer functions which yield the probability for an observed lepton to carry a given fraction of the τ lepton energy.

In the reconstruction of the final state, the transverse momentum of the Zjj system is assumed to be zero. While developing the method two additional procedures have been tested. In the first one, the transverse momentum of the Zjj system was not constrained at all, while in the second one, the final probability was multiplied with the a priori probability for the corresponding transverse momentum. As in the case of the top pair momentum, this probability was derived using Monte Carlo simulated events. Since the assumption of zero transverse momentum yields the best separation of signal and background, the computation is significantly faster, and the effect is less important than in the calculation of the signal likelihood, this assumption is used to calculate the background likelihood.

To assess the $(Z \rightarrow \tau\tau)\,jj$ likelihood, the directions of the τ leptons are assumed to be identical to those of the final-state electron and muon, which is a good approximation because of the large boost of the τ leptons from Z decay. The jet and lepton directions are directly taken from the detector measurements, while the two jet energies are obtained from random numbers distributed according to jet transfer functions under the assumption of light-flavor jets. For each event, at least 10,000 random numbers are drawn. If needed, up to a maximum of 50,000 are drawn, until the standard deviation of the individual values is no larger than 10% of the average. The absolute values of the two τ momenta are determined from the constraint of zero event transverse momentum, $p_T^{Zjj} = 0$. Then, the sum of the matrix elements times PDF factors is calculated as in the $(Z \rightarrow \ell\ell)jj$ case using VECBOS, and multiplied with the phase space factor and the jet transfer function factors. To account for the τ lepton decay, the τ transfer functions are included yielding the likelihood to observe the lepton energy for the given τ energy.

---

[2] Heavy-flavor quarks are not considered separately as the kinematics of these events is very similar and no bias is observed, see Fig. 6.12.

5.8 Normalization and Performance of the Background Likelihood

In principle, the normalization of the background likelihoods can be calculated in the same way as the signal likelihoods are normalized. However, the computation of the integral given in Eq. 5.9 would be very computing intensive, and a different approach is chosen. It makes use of the fact that only the correct background-to-signal normalization reproduces the correct signal fraction. Thus, the following procedure is used to determine the relative background normalization.

- For each channel and each top mass, one large ensemble has been built using $t\bar{t}$ and $(Z \to \tau\tau)$ + jets Monte Carlo simulated events.
- The signal fraction is fitted in the sample and the normalization of the background likelihood is adjusted, until the estimate yields the true signal fraction within $\varepsilon = 0.005$.
- Since the normalization of the background likelihood cannot depend on the top quark mass, the above steps are applied to each available $t\bar{t}$ Monte Carlo sample. The mean of all results is taken as the signal to background normalization scale.

To demonstrate the good performance of the background likelihoods, Fig. 5.12 shows the separation of $t\bar{t}$ signal and $(Z \to \tau\tau)$ + jets background events in the electron + muon channel using signal and background likelihoods. The top mass hypotheses of the signal likelihoods have been chosen such that they correspond to the generated top masses of 160 GeV in the upper, 170 GeV in the middle and 185 GeV in the lower plot. Both signal and background likelihoods are normalized as described above. To make use of signal events where the requirement of zero Zjj transverse momentum cannot be fulfilled, the background likelihoods are set to 1×10^{-31}. This explains the vertical band in the plots. 4% of the $(Z \to \tau\tau)$ + jets events and 40% of the signal events fail this requirement.

5.9 Likelihood Evaluation

As already noted above, the Matrix Element method can be extended to measure any quantity the event likelihood depends on. Even though the calculation of the event likelihood would be more or less the same, the extraction of the observable could be quite different. Thus, the present section will explicitly focus on the measurement of the top quark mass. Required modifications will be discussed in Sect. 7.2.3. The calculation of the event likelihood includes the following steps:

- For each signal and background event, the signal likelihood is calculated as a function of the top mass hypotheses and the two jet-parton assignments using the two leading jets in each event. The step width of the mass hypotheses is chosen such that it is about half the expected statistical error, i.e. 2.5 GeV.

Fig. 5.12 Normalized signal likelihoods compared to normalized background likelihoods for both signal $t\bar{t}$ and background $(Z \to \tau\tau)$ + jets events. The mass hypothesis corresponds to the generated mass of the signal sample; top: 160 GeV; middle: 170 GeV; bottom: 185 GeV. The lines correspond to $L_{t\bar{t}} = L_{(Z \to \tau\tau)jj}$.

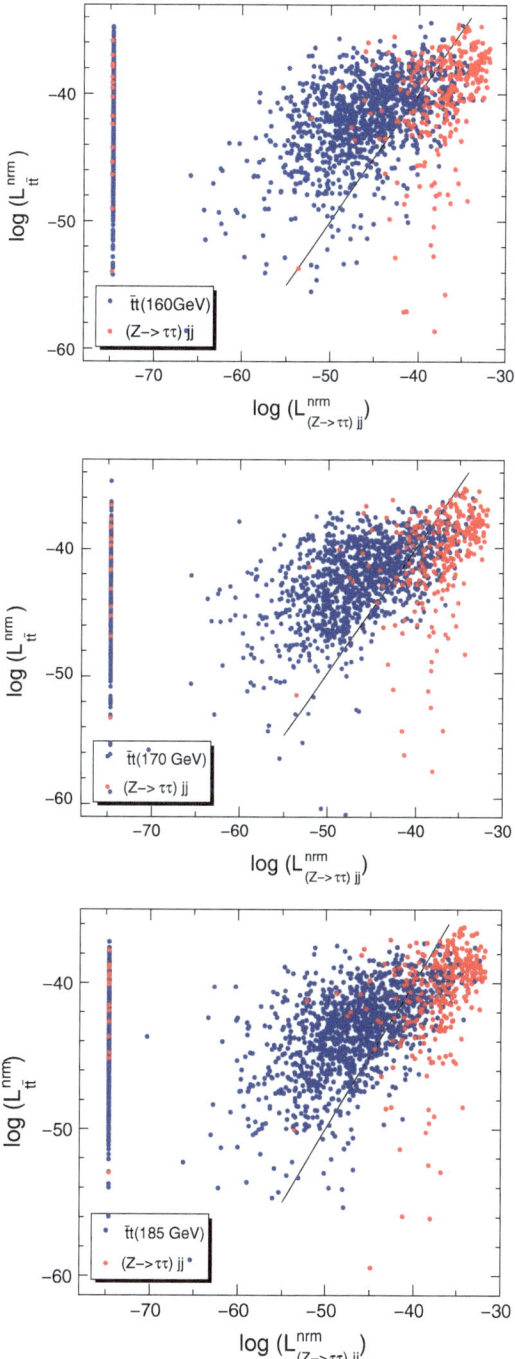

- The mean of the two jet-parton assignments is computed for each hypothesis and the signal likelihood is normalized according to Eq. 5.26 taking the selection cuts into account.
- Next, the background likelihood is calculated for the two jet-parton assignments using the two leading jets.
- As for the signal likelihood, the mean of the two jet-parton assignments is computed and the background likelihood is normalized with the relative background-to-signal scale, discussed in Sect. 5.7.
- For each event, the signal and background likelihoods are combined to the event likelihoods according to Eq. 5.1.
- In a last step, all event likelihoods are combined to the sample likelihood as described in Eq. 5.3.

The extraction of the top quark mass from the sample likelihood is done as follows:

- First, the global maximum of the sample likelihood (5.3) is determined with respect to the signal fraction, $f_{t\bar{t}}$, by scanning the signal fraction in steps of 0.004 from 0 to 1 for all mass hypotheses. The error on the signal fraction, $f_{t\bar{t}}^{best}$, follows from the difference to the point where the logarithm of the sample likelihood is shifted by 0.5. Thus, no correlation between the signal fraction and the top mass is accounted for in this uncertainty.
- With the fitted signal fraction, $f_{t\bar{t}}^{best}$, the sample likelihood is now reduced to a function that depends only on the parameter to be measured, the top quark mass.
- The top mass is measured as the minimum of a polynomial fit of the 2nd-order to the negative logarithm of the sample likelihood, which corresponds to the maximum of a Gaussian fit to the sample likelihood. The fit range is chosen such that three top mass hypotheses of either side around the minimum are included.
- The statistical uncertainty on the top quark mass is given by the 68% confidence region around the measured value.

5.10 Ensemble Testing

An important tool in the measurement of the top quark mass is the so-called ensemble testing procedure allowing for a validation and calibration of the method, and a precise estimate of the statistical error in the event of low statistics. Especially in case of the Matrix Element method, the ensemble testing is indispensable, since the calculation of the event likelihoods is very CPU intensive and only a relatively small number of Monte Carlo simulated events can be used. The computation of the signal probability for one top mass hypothesis and 1,000 events employs one Intel Xeon 2.3 GHz CPU for 160–300 h, depending on the sampling rate of the numerical integration.

In the ensemble-testing procedure, multiple pseudo-experiments are performed in analogy to the final measurement in data using Monte Carlo simulated events. Each of the pseudo-experiments yields a measurement of the top quark mass, its statistical uncertainty, and its pull, i.e. the deviation of the measurement from the true top mass in units of the statistical uncertainty. Each ensemble is composed according to the measured composition of the data set where the fractions of the various samples are allowed to fluctuate following Poisson statistics.

In the case of weighted Monte Carlo events, the weights are taken into account when building the ensembles. Technically, a random number is drawn in the range of the weights and only in case the weight exceeds the random number, this event is added to the ensemble. In the fit however, all events in the pseudo-experiment are treated as unweighted, i.e. the same way as it is done for data.

Pseudo-experiments are built such that events are not removed from the pool of available events once they are drawn. Each event is allowed to appear in multiple pseudo-experiments, and even multiple times in the same pseudo-experiment. The precision on the statistical uncertainty increases significantly when using the technique of resampling [18, 19]. Since the number of pseudo-experiments (1,000) is sufficiently larger than the maximum number of independent pseudo-experiments (30–40), no bias is expected from the fact that some events are used more often than others. The number of ensembles is chosen such that no further information on the measurement of the top mass can be gained by increasing the number of pseudo-experiments.

To avoid any bias from events where the calculated mass probabilities span an unreasonable range of more than 20 orders of magnitude, a cut against these events is applied during the ensemble building. The efficiency of this cut is 100% for parton-matched events and more then 99.5% for fully simulated Monte Carlo events.

As already mentioned above, the technique of ensemble testing is used to validate and to calibrate the Matrix Element method. The derivation of a calibration curve includes three steps:

- Using background and signal events of a given top mass e.g. 170 GeV, a pseudo-experiment is built according to the measured composition in data. The top mass as well as its uncertainty and pull are determined following the description in Sect. 5.9.
- The first step is repeated for 1,000 pseudo-experiments. The resulting top masses, their errors and pulls are combined in one plot to fit the mean top mass and to determine the mean statistical uncertainty and pull width. The latter is a measure for the validity of the statistical uncertainty. If the statistical uncertainty were underestimated, the pull width would be larger than 1 and the statistical uncertainty would need to be corrected for this.
- In a last step, ensemble tests are performed for pseudo-experiments with signal events of additional generated top mass points to derive a calibration curve for the top mass measurement. The curve relates the true top masses, m_{top}^{cal}, to the measured ones, m_{top}^{raw}, and allows to validate and calibrate the method.

The calibration curve is parametrized as a linear function around the central top mass, $m_{\text{top}}^{\text{cent}}$, given by

$$m_{\text{top}}^{\text{raw}} = s\left(m_{\text{top}}^{\text{cal}} - m_{\text{top}}^{\text{cent}}\right) + o + m_{\text{top}}^{\text{cent}}, \qquad (5.36)$$

where o denotes the offset of the calibration curve with respect to the central top mass and s the slope. The slope and offset obtained are used to correct the raw measurement.

- When calculating the uncertainties on the calibration of the top mass, the resampling has to be taken into account [19]. For the mean of the top mass and its uncertainty, the error is given by the width of the distribution divided by the number of the independent ensembles, so

$$\sigma_{m_{\text{top}}} \Rightarrow \sigma_{m_{\text{top}}} \sqrt{\frac{n_{\text{evts}}}{n_{\text{pool}}}} \qquad (5.37)$$

$$\sigma_{\Delta m_{\text{top}}} \Rightarrow \sigma_{\Delta m_{\text{top}}} \sqrt{\frac{n_{\text{evts}}}{n_{\text{pool}}}}. \qquad (5.38)$$

The error on the width of the pull distribution is given by

$$\sigma_{m_{\text{top}}\text{pull width}} \Rightarrow \sigma_{m_{\text{top}}\text{pull width}} \sqrt{\frac{1}{2}\left(\frac{1}{n_{\text{pool}}} + \frac{1}{n_{\text{ens}} - 1}\right)}, \qquad (5.39)$$

according to Mulders [19] where n_{pool} is the number of events in the pool, n_{evts}, the number of events in one ensemble, and n_{ens}, the number of pseudo-experiments.

References

1. Haefner P et al (2005) Determination of the muon transfer function for top mass measurements. DØ note 4818
2. Schieferdecker P (2005) Measurement of the top quark mass at DØ Run II with the Matrix Element method in the lepton + jets final state. Dissertationsschrift, Ludwig-Maximilians-Universität München
3. Wang   M.   http://www-d0.hef.kun.nl//askArchive.php?base=agenda&categ=a071376&id= a071376s1t3/transparencies/mestat_jetmass_16aug07.pdf
4. Ochando C. http://www-d0.hef.kun.nl//askArchive.php?base=agenda&categ=a08924&id= a08924s1t3/transparencies
5. Gutierrez G et al (2008) Measurement of the top quark mass in the lepton + jets channel using the Matrix Element method on 2.2 fb^{-1} of DØ Run II data. DØ note 5750
6. James F. MINUIT reference manual, CERN Program library writeup D506
7. Grohsjean A (2008) Jet transfer function determination using top quark MC with final Run IIa JES and improved JSSR. DØ note 5770
8. Grohsjean A, Fiedler F (2008) Measurement of the top quark mass with the Matrix Element method in the dilepton channel. DØ note 5640
9. Abazov VM et al (DØ Collaboration) (2006) Measurement of the top quark mass in the lepton + jets final state with the Matrix Element method. Phys Rev D 74:092005

10. Biebel O (2001) Experimental tests of the strong interaction and its energy dependence in electron–positron annihilation. Phys Rep 340:165
11. Fiedler F (2007) Precision measurements of the top quark mass. Habilitationsschrift, Ludwig-Maximilians-Universität München
12. Mahlon G, Parke S (1997) Maximizing spin correlations in top quark pair production at the Tevatron. Phys Rev Lett 80:2063
13. Yao WM et al (2006) Review of particle physics. J Phys G 33:1
14. Lepage GP (1978) A new algorithm for adaptive multi-dimensional integration. J Comput Phys 27:192
15. Lepage GP (1980) Vegas: an adaptive multi-dimensional integration program. Cornell preprint CLNS:80–447
16. Galassi M et al GNU scientific library reference manual (2nd ed.). ISBN 0954161734. http://www.gnu.org/software/gsl/
17. Berends FA et al (1991) On the production of a W and jets at hadron colliders. Nucl Phys B 357:32
18. Barlow R (2000) Application of the bootstrap resampling technique to particle physics experiments. MAN/HEP/99/4
19. Mulders M (2004) Ensemble testing for the top mass measurement. DØ note 4460

Chapter 6
Measurement of the Top Quark Mass

The following chapter describes the first measurement of the top quark mass with the Matrix Element method in the dilepton channel at the DØ experiment. The chapter starts with a detailed description of the data samples and the event selection. Afterwards, in Sect. 6.2, the method is validated using so-called parton-level events. These events are not run through the full detector simulation but step by step, the transfer functions being applied to simulate different detector effects. This procedure allows a bias-free implementation and a performance check of the method. Before the method is finally applied to the selected data set in Sect. 6.4, a calibration curve is derived in Sect. 6.3 using Monte Carlo generated events which are processed with the full detector simulation. The systematic uncertainties, accounting for differences between the Monte Carlo simulated events and data, are discussed in Sect. 6.5. The chapter ends with the combination of the presented results.

6.1 Data Samples and Event Selection

The topology of the dilepton decay channel shown in Fig. 6.1 is characterized by two energetic b jets, arising from the two b quarks, and two isolated leptons, one electron and one muon, of opposite charge. In addition, there is a significant amount of missing transverse energy due to the undetected neutrinos. Accounting for the event topology, dilepton $t\bar{t}$ candidate events are triggered by the presence of high-p_T muons or high-E_T electrons. The main physics background with a similar signature in the detector comes from $(Z \to \tau\tau)$ + jets events where one τ lepton decays into an electron, and the other one, into a muon. Since the jets arise from soft radiation and the measured leptons only carry part of the energy from the τ leptons, this background can be significantly reduced requiring a certain amount of transverse momentum for the scalar sum of the two leading jets and the leading lepton. Additional background comes from WW + jets events where the W boson is

A. Grohsjean, *Measurement of the Top Quark Mass in the Dilepton Final State Using the Matrix Element Method*, Springer Theses,
DOI: 10.1007/978-3-642-14070-9_6, © Springer-Verlag Berlin Heidelberg 2010

Fig. 6.1 Schematic top
quark decay in the
electron + muon final state

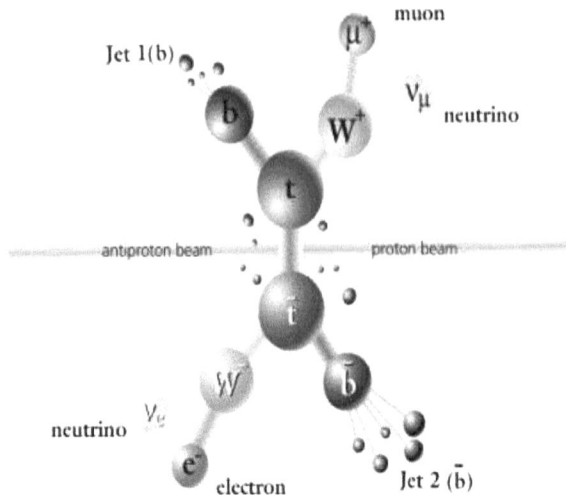

produced in association with jets. Since the cross section for this process is about two orders of magnitude smaller than the one for $(Z \rightarrow \tau\tau)$ + jets events and since the jets also arise from soft radiation, the same cut can be used to reduce this kind of background. The next source of background are WZ + jets events where one of the electrons produced is misidentified as a jet. The cross section for this process is again one order of magnitude smaller. WW + jets and WZ + jets events are considered together and referred to as diboson events. The main instrumental background in this channel arises from events with one leptonic W decay and at least three jets, one of which is misidentified as another isolated lepton. Jets can appear as isolated electrons if they contain a pion that decays into two photons leading to a large amount of electromagnetic energy in the calorimeter. Additional contributions come from the semileptonic decay of c and b hadrons inside jets. The instrumental background can be reduced by requiring the leptons to be of good quality and to have opposite charge. In addition, they must be spatially isolated from jets.

Two different data sets are used to measure the top quark mass in the present analysis, the Run IIa and Run IIb data sets. The first one was recorded between April 2002 and February 2006 and corresponds to an integrated luminosity of $1.1 \, \text{fb}^{-1}$. The second was recorded between June 2006 and May 2008 and corresponds to an integrated luminosity of $1.7 \, \text{fb}^{-1}$. As already discussed in Sect. 2.2, the DØ detector was upgraded between these two run periods so that both data samples need to be analyzed separately. However, for both periods almost the same selection is applied.

To study the data samples selected, they are compared to samples of Monte Carlo simulation. The $t\bar{t}$ signal events are generated according to the matched ALPGEN + PYTHIA scheme. This means ALPGEN [1] generates the contribution from the parton-level matrix elements, while PYTHIA [2] performs the evolution of the parton shower and the hadronization of partons to hadrons. The MLM matching scheme [3] is applied afterwards. It removes the possible overlap between the hard partons created by the matrix elements and the parton shower. To reproduce the full $t\bar{t}$ spectrum, three different types of samples are used: $t\bar{t} + 0$lp, $t\bar{t} + 1$lp, $t\bar{t} + 2$lp. In the so-called 0lp and 1lp bin, exactly 0 and 1 light parton are generated in association with the pure $t\bar{t}$ process, whereas in the 2lp bin 2 or more light partons are generated additionally. The signal $t\bar{t}$ samples are produced with the leading-order parton distribution function CTEQ6L1 [4]. The events in the central sample, used for the data-to-Monte Carlo comparisons, are weighted to reproduce the cross section calculated in next-to-leading order by N. Kidonakis, 7.91 ± 0.50 pb [5].

The simulated $(Z \to \tau\tau) +$ jets events are generated with ALPGEN + PYTHIA using the leading-order parton density function CTEQ6L1. The next-to-next-to-leading-order cross section is calculated to be $256.6^{+5.1}_{-12.0}$ pb [6] and events in the sample are weighted accordingly. As the transverse momentum of the Z boson is not properly described in ALPGEN, the distribution of the transverse momentum is reweighted for different jet bins [7].

The $WW +$ jets and $WZ +$ jets background samples are produced using PYTHIA and the leading-order parton density function CTEQ6L1. Both samples are normalized to the next-to-leading-order cross section which is about 40% higher than the leading-order cross section given by PYTHIA [8].

All Monte Carlo generated events are processed with the full GEANT detector simulation. The same reconstruction as in case of the data is used. A summary of the Monte Carlo samples is given in Table 6.1.

The event selection used in this measurement is optimized with respect to two different goals. First of all, to achieve a good agreement between measured data and Monte Carlo simulated events. Secondly, to reduce the number of background events contaminating the selected sample. In the following, the details of the selection are discussed.

- *Data Quality* The standard data quality selection described in Sect. 3.9 is applied, i.e. bad luminosity blocks are removed, and all calorimeter-event-quality variables are required to be good.
- *Monte Carlo Corrections* As the luminosity profile is different between the zero-bias events in data and the zero-bias events added to the Monte Carlo samples, the Monte Carlo events are reweighted accordingly [9]. In addition, a weight is applied to correct for the distribution of the primary vertex position along the beam axis [10].

Table 6.1 Summary of Monte Carlo samples and the corresponding cross sections σ used for the data-to-Monte Carlo comparisons

Process	Generator	σ (pb)
$t\bar{t}$, $m_{\text{top}} = 170$ GeV	ALPGEN + PYTHIA	7.91
$(Z \to \tau\tau) +$ jets, $m_{\tau\tau} = 15 - 60(75)$ GeV	ALPGEN + PYTHIA	409 (424)
$(Z \to \tau\tau) +$ jets, $m_{\tau\tau} = 60(75) - 130$ GeV	ALPGEN + PYTHIA	256 (241)
$WW +$ jets	PYTHIA	1.26
$WZ +$ jets	PYTHIA	0.12

The values in brackets refer to the ones different for Run IIb

- *Muons* The muon and the matched track are required to be of medium quality [11]. Additionally, the muon must be detected in all three layers of the muon system. Besides the isolation criteria described in Sect. 3.4, muons need to have a minimum transverse momentum of 15 GeV. Cosmic muons are removed by a timing requirement as described in Sect. 3.4. As already discussed in Sect. 5.3.2, the shutdown of the DØ detector in fall 2004 has led to a change of the muon momentum resolution in Run IIa. Thus, two different smearings are applied to the Run IIa muons which are referred to as pre- and post-shutdown smearing. 40% of the Monte Carlo simulated events for Run IIa are considered to be pre-shutdown events. To account for discrepancies in the selection efficiency between data and Monte Carlo simulated events, an additional correction factor is applied [11].
- *Electrons* Besides the quality cuts, discussed in Sect. 3.3, electrons are required to have a minimum transverse momentum of 15 GeV. To remove the background from muon bremsstrahlung, electrons are not allowed to have a common track with any muon matched to a loose track or any loose muon. Events with additional electrons are removed. As in the case of muons, a correction factor is applied to account for discrepancies in the selection efficiency between data and Monte Carlo events [12].
- *Jets* Each event is required to have at least two jets. The leading jet must have a minimum transverse momentum of 30 GeV after all jet corrections, the trailing jet, of 20 GeV. For the Run IIb measurement, the cut on the leading jet is also lowered to 20 GeV to increase the statistics. Additionally, jets must fulfill the quality criteria described in Sect. 3.6. Jets with a distance of $\Delta R < 0.5$ to an electron are removed.
- *Vertex Selection* The primary collision vertex is required to have a maximum distance of 60 cm to the central point of the detector. Moreover, at least 3 tracks must be associated with the primary vertex and the distance between the electron or muon selected and the primary vertex must be less than 1 cm.
- H_T The sum of the transverse momenta of the two jets and the leading lepton must be at least 115 GeV. This requirement is applied to reduce the dominant background of $(Z \rightarrow \tau\tau) +$ jets events.

Two different kinds of instrumental background are distinguished: fake electrons (fake e) and fake isolated muons (fake iso. μ). Fake electrons refer to events where a jet is misidentified as an electron, as well as to events with real electrons produced in jets. Since these electrons are not isolated and their electron-likelihood value, see Sect. 3.3, is low, both types of events are denoted as fake electrons. To determine the fraction of fake electrons in the $t\bar{t}$ data sample, the shape of the electron-likelihood distribution is measured in a pure $Z \rightarrow ee$ sample and in a multijet and $(W \rightarrow \mu\nu_\mu) +$ jets dominated background sample, see Fig. 6.2. Both distributions are compared to the electron likelihood in the data sample and the fraction of fake electrons is extracted.

The fake isolated muons mainly arise from events with one isolated electron and one fake isolated muon from heavy-flavor-quark decay, e.g. $(W \rightarrow e\nu_e) +$ jets

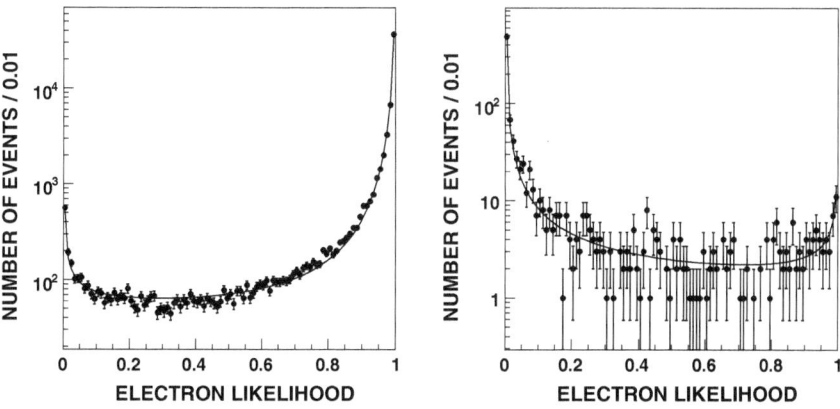

Fig. 6.2 Electron likelihood distributions; *left* for real electrons from $Z \rightarrow ee$ data events; *right* for fake electrons from a multijet and $(W \rightarrow \mu\nu_\mu)$ + jets dominated background sample [8]

Table 6.2 Final number of expected and observed events and their statistical uncertainties after all cuts for Run IIa and Run IIb

	$t\bar{t}$	$(Z \rightarrow \tau\tau)$ + jets	WW + jets, WZ + jets	Fake e	Fake iso. μ	Data
Run IIa	$36.77^{+2.59}_{-2.69}$	$5.95^{+0.88}_{-0.99}$	$1.61^{+0.40}_{-0.39}$	$0.75^{+0.26}_{-0.24}$	$1.85^{+0.50}_{-0.45}$	39
Run IIb	$50.15^{+2.53}_{-2.74}$	$6.88^{+0.76}_{-0.96}$	$2.41^{+0.50}_{-0.50}$	$1.20^{+0.52}_{-0.45}$	$3.25^{+0.90}_{-0.80}$	68

and $(Z \rightarrow ee)$ + jets events. The number of fake isolated muons in the data sample is estimated as the product of the fake muon isolation probability and the number of events with two leptons of same charge, where the muon is not required to be isolated and the fake electron background is subtracted. To evaluate the fake muon isolation, a non-isolated muon in $b\bar{b}$ data events is selected, and the rate of the second muon to appear isolated is measured.

As by now, no proper Monte Carlo samples are available to use instrumental background in the ensemble-testing procedure, this background can only be treated as a systematic uncertainty in the measurement of the top quark mass.

The final number of expected and observed events after all selection and quality cuts are summarized in Table 6.2. The data-to-Monte Carlo comparison plots for the main variables used in the calculation of the event likelihood are shown in Figs. 6.3, 6.4, 6.5, 6.6 for both Run IIa and Run IIb. All data-to-Monte Carlo comparisons show a good agreement for the kinematic distributions, and no large discrepancies are observed in any of them. More details on the selection are given by M. Arthaud [8, 13].

6.2 Parton-Level Studies

Since the Matrix Element method is used for the first time to measure the top quark mass in the dilepton final state at the DØ experiment, fundamental tests are carried

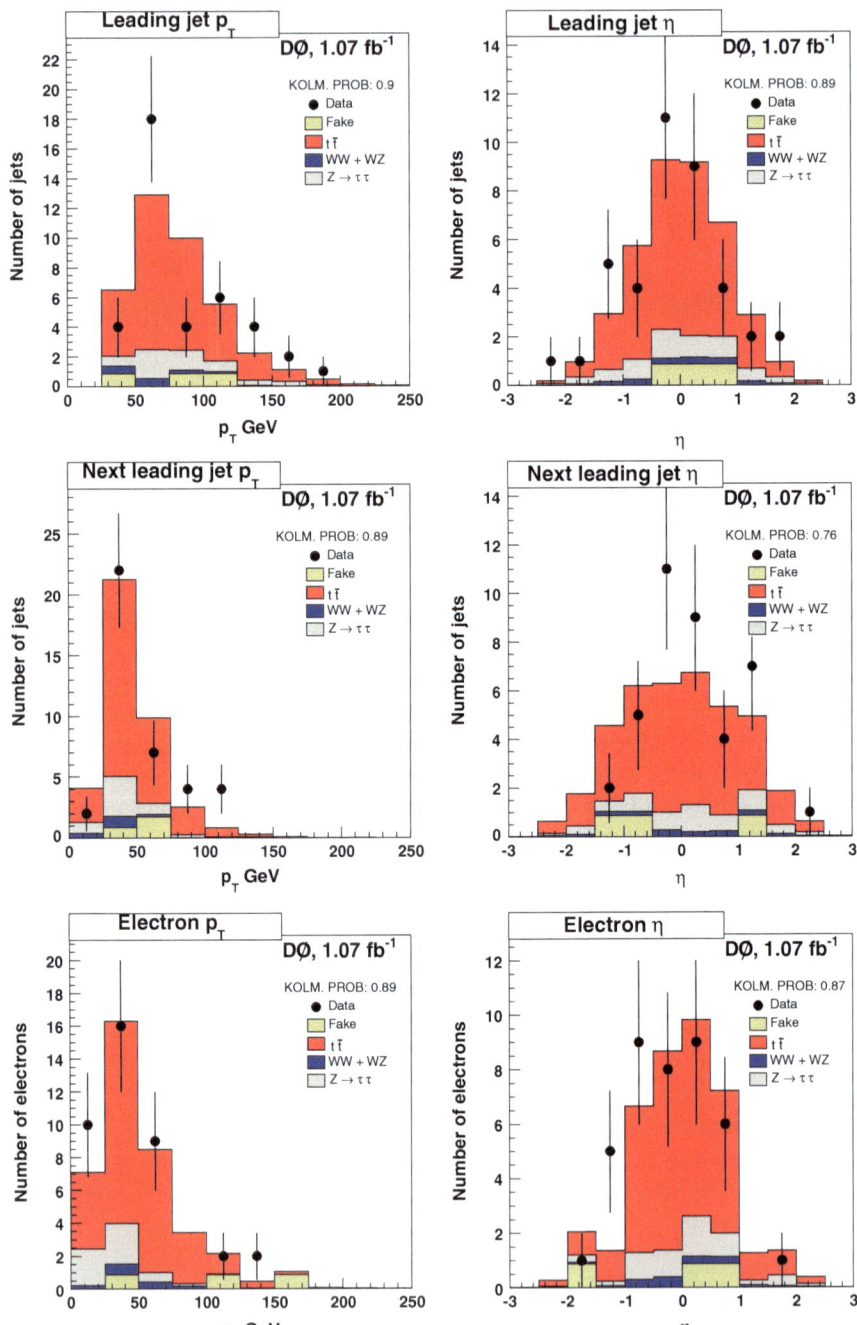

Fig. 6.3 Data-to-Monte Carlo comparisons of the transverse momentum and the pseudorapidity of the two leading jets and the electron for Run IIa after all cuts [8]

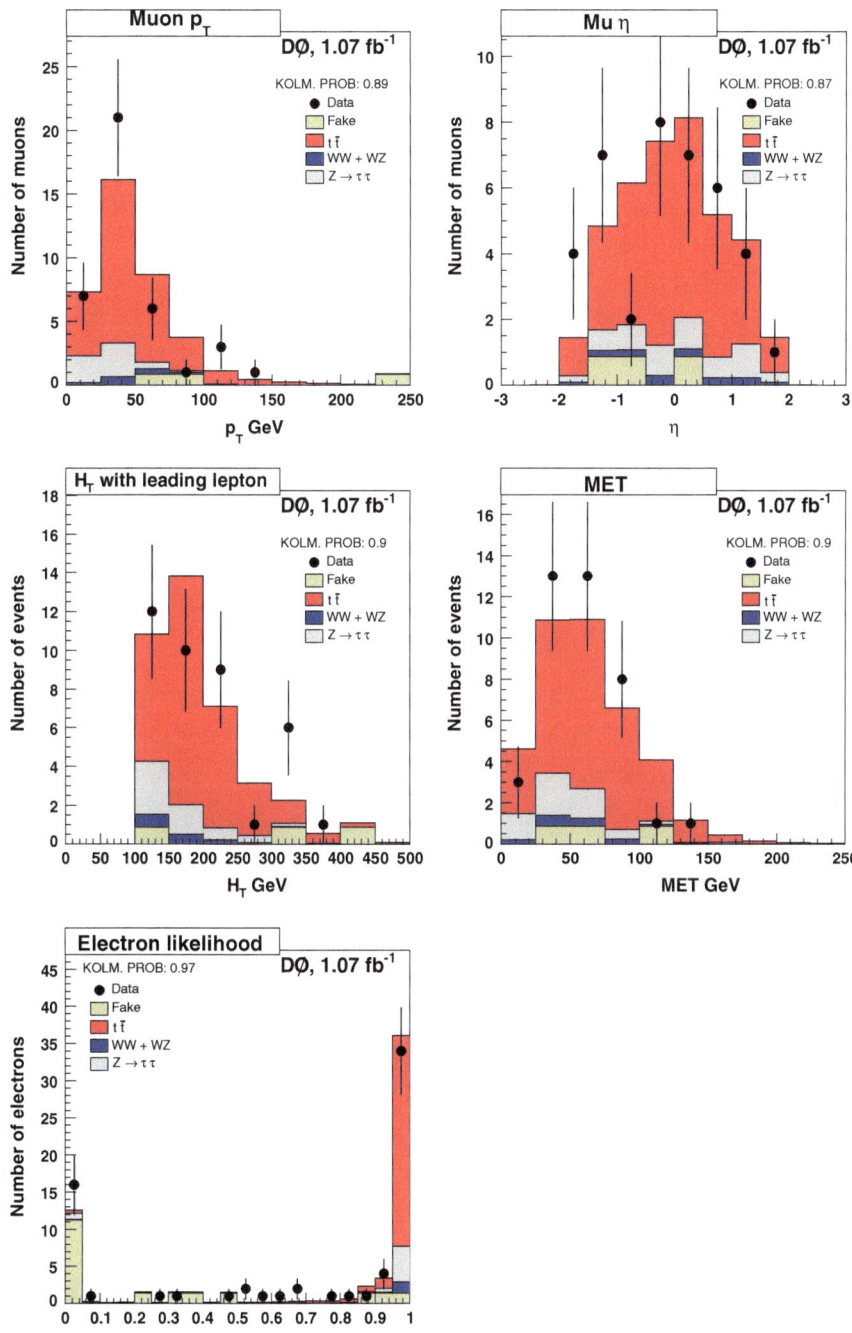

Fig. 6.4 Data-to-Monte Carlo comparisons of the transverse momentum and the pseudorapidity of the muon, the missing transverse energy, H_T and the electron likelihood for Run IIa after all cuts [8]. In case of the electron likelihood, the cut on the likelihood is not included

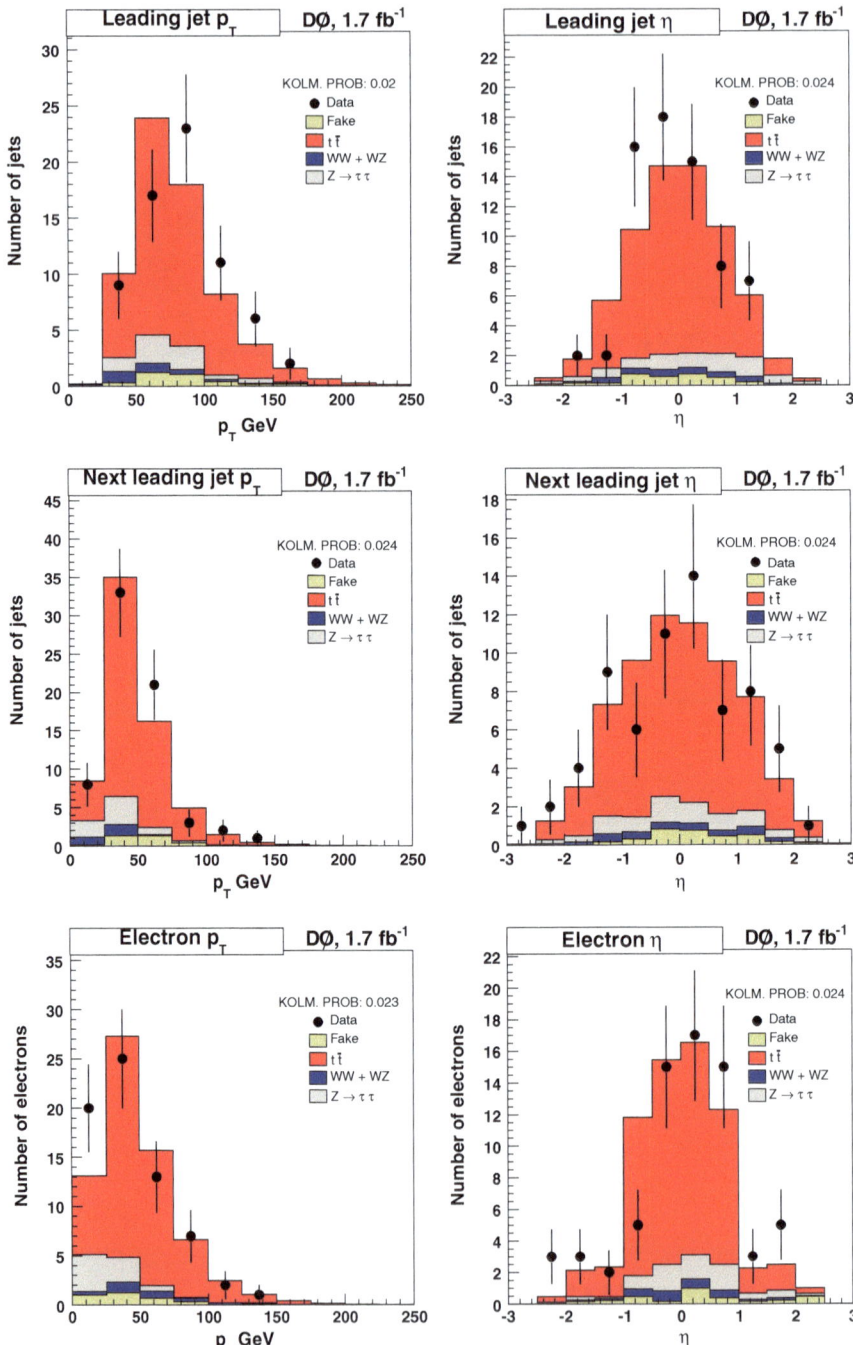

Fig. 6.5 Data-to-Monte Carlo comparisons of the transverse momentum and the pseudorapidity of the two leading jets and the electron for Run IIb after all cuts [13]

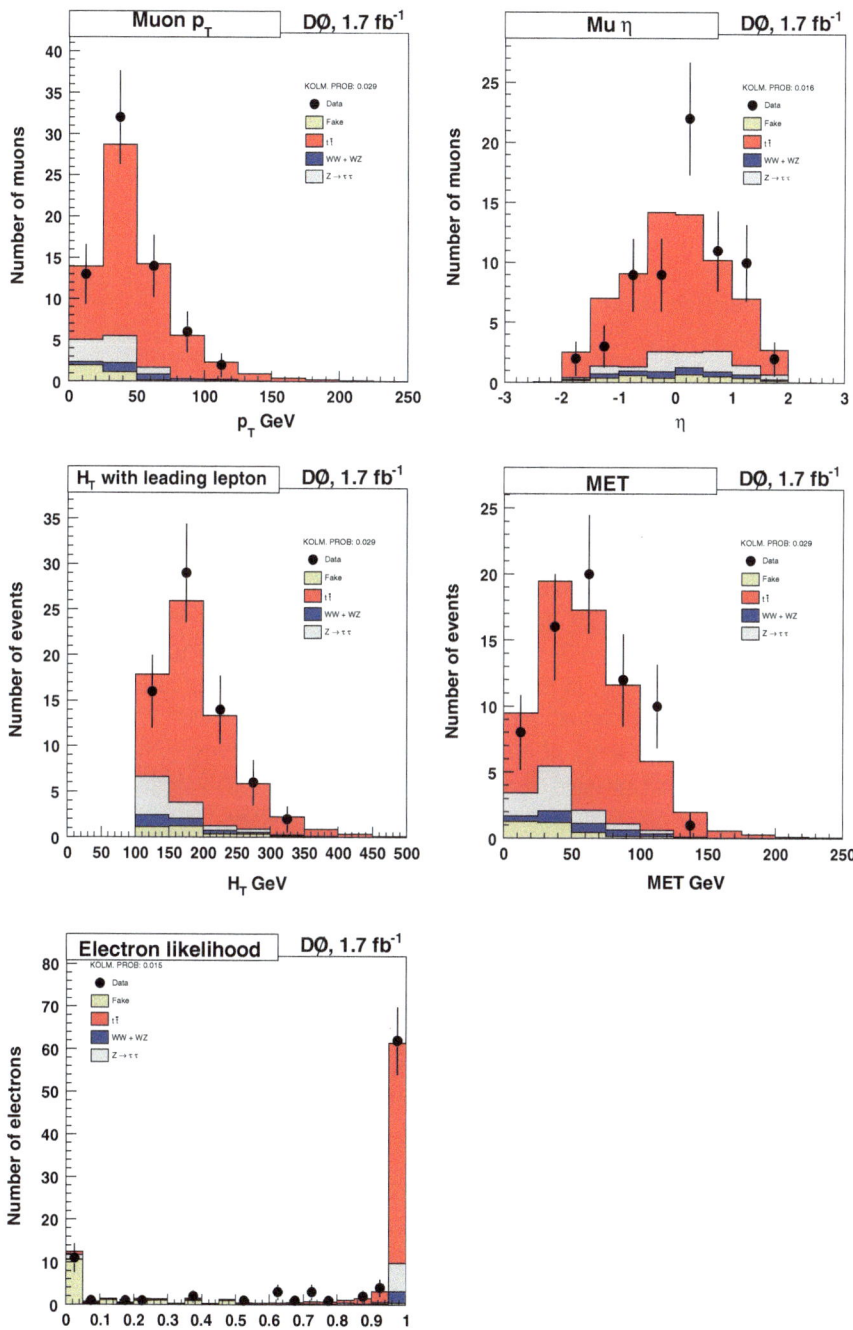

Fig. 6.6 Data-to-Monte Carlo comparisons of the transverse momentum and the pseudorapidity of the muon, the missing transverse energy, H_T and the electron likelihood for Run IIb after all cuts [13]. In case of the electron likelihood, the cut on the likelihood is not included

out to validate the method and check its performance. This is especially important as the kinematic reconstruction of the partons required a complete revision of the procedure used in the semileptonic analysis. Moreover, the dilepton channel demanded the design of a new background likelihood modeling the decay of a τ lepton.

To allow for an idealized test bed, the studies presented in this section are based on so-called parton-level events, which are simulated with a leading-order-matrix-element generator and smeared according to the measured transfer functions. They are not run through the full detector simulation and fulfill the assumptions made for the calculation of the event likelihoods.

6.2.1 Monte Carlo Samples

The Monte Carlo samples used to validate the method are produced with leading-order Monte Carlo generators including no initial- or final-state radiation. The $t\bar{t}$ signal events are generated with MADGRAPH [14], the Zjj and $WWjj$ background events, with ALPGEN [1]. For the modeling of the parton distribution functions, the leading-order PDF CTEQ5L [4] is chosen.

To simulate the decay of a τ lepton to an electron or muon in the Zjj events, the τ transfer function shown in Fig. 5.9 in Chap. 5 is applied, while the direction of the lepton remains unchanged. As described in Sect. 5.3.3 in Chap. 5, the τ transfer function models the energy of a lepton coming from a τ decay.

The signal and background events generated are not run through the full detector simulation, but the parton momenta are smeared according to the transfer functions discussed in Sect. 5.3 in Chap. 5. The resolution represents the real DØ detector well while it is not subject to any uncertainty. As assumed by the method, the direction of the particles stays the same. This allows for a one-to-one mapping between the partons at tree-level and the final-state particles, whereas in fully reconstructed events, jets may for example arise from physics or detector effects such as radiation or calorimeter noise. Thus, the kinematic spectra of the parton-level events correspond well to the leading-order matrix element used in the likelihood calculation. As the background samples contain only light partons, b jets are simulated by smearing the light partons with the transfer functions for b jets.

To allow for a measurement of the top quark mass at different mass points, samples are generated at 160, 165, 170, 175, and 180 GeV. As the computation of the event likelihoods is very CPU intensive, only a limited number of Monte Carlo events is used, e.g. 2000 $t\bar{t}$ events per mass point. However, about 30–40 independent samples can be built, and it is checked that no bias is expected due to the resampling.

To minimize computing time, most of the parton-level studies are performed assuming perfectly measured lepton momenta, i.e. the generated leptons are not smeared, and the additional integrations over the inverse lepton momenta are not carried out. It is verified that the conclusion from the parton-level tests stays the same and is not affected by this choice. To validate the integration over the inverse muon momentum, an additional test is performed using smeared leptons.

The simulated events are required to pass a simplified selection including the following cuts:

- $p_T^j > 20$ GeV
- $|\eta_j| < 2.5$
- $p_T^\ell > 15$ GeV
- $|\eta_e| < 1.1$ or $1.5 < |\eta_e| < 2.5$
- $|\eta_\mu| < 2.0$
- $\not{E}_T > 30$ GeV
- $\Delta R(j, j') > 1.0$
- $\Delta R(\ell, j) > 0.5$

where j denotes a jet, ℓ a lepton, e an electron, and μ a muon. The missing transverse energy, $\not{E}_T$, is calculated according to Eq. 5.34. The ΔR cut between two jets, and a jet and a lepton, corresponds to the minimal distance to a jet[1] which is restricted by the geometrical jet cone size of 0.5.

6.2.2 Normalization

The normalization of the signal likelihood follows the description given in Sect. 5.6, and the selection cuts listed above are taken into account. As for the generation of the parton-level events, the leading-order parton distribution functions in the version CTEQ5L are used. The calculated normalization as a function of the top quark mass is fitted with a 3rd-order polynomial.

As an example, Fig. 6.7 shows the normalization curve for the electron + muon parton-level selection. The additional integration over the inverse lepton momentum is carried out to account for the uncertain muon resolution. The parameters of the fit are listed in the upper right corner of Fig. 6.7.

The normalization of the background likelihood is calculated following the description in Sect. 5.8. The relative background-to-signal scale depends on two parameters: the signal fraction and the top quark mass. The first dependency can be traced back to the imperfect modeling of the background likelihoods as they are evaluated with VECBOS, while the events are generated with ALPGEN. In addition, several approximations are made, like the modeling of the τ lepton decay by the τ transfer functions. The dependency on the top quark mass is caused by the similarity of the background events to the signal events at small top quark masses. In this analysis, the signal fraction is chosen to correspond to the expected fraction in data,

---

[1] In fact, this does not hold for fully reconstructed jets as they are subject to splitting and merging, and the minimum distance between two jets can be 0.5.

Fig. 6.7 Observable cross section for the electron + muon parton-level selection. The parameters from the fit of a 3rd-order polynomial are given in the *upper right box*. The branching ratios of the top quark decay are not included as constant factors are irrelevant for the mass measurement

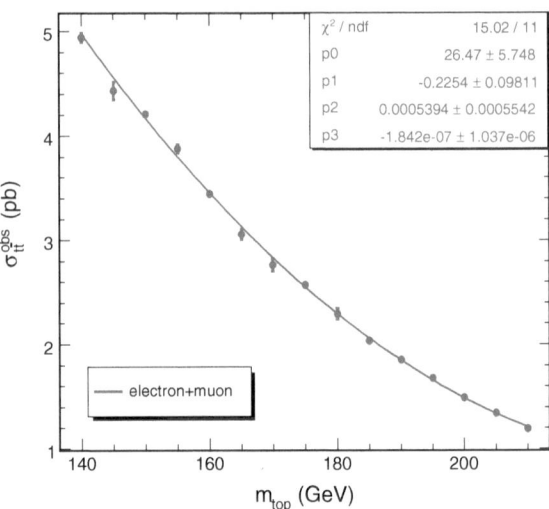

i.e. 80%, and the relative scale is derived as a function of the generated top mass using $t\bar{t}$ and $(Z \rightarrow \tau\tau)\,jj$ events. To minimize statistical fluctuations, the ensemble size is maximized with respect to the available pool size. Thus, $1,200\ t\bar{t}$ and $300\,(Z \rightarrow \tau\tau)\,jj$ events are used. As no mass dependency can be considered, the mean of all values is taken as the relative background-to-signal scale. However, varying the relative scale within its statistical uncertainties has no effect on the measured top quark mass.

6.2.3 Signal-Only Studies

In a first step, tests are performed using ensembles of pure signal events. The partonic jet energies are smeared following the b jet transfer functions; the lepton momenta are assumed to be perfectly measured. Accordingly, the integration over the muon transverse momentum is fixed and the background likelihoods are not included. Like for all parton-level studies, the integration over the top pair transverse momentum is not carried out as the top pair system in the parton-level events is produced at rest in the transverse plane. For the simulated top quark masses of 160, 165, 170, 175, and 180 GeV, a total of 1,000 pseudo-experiments is performed per mass point. Each of the ensembles is built of 50 $t\bar{t}$ events.

The results from the 1,000 pseudo-experiments are shown on the left side of Fig. 6.8. The upper plot depicts the measured top quark masses, the middle, their uncertainties and the lower, their pull, i.e. the deviation of the mass from the true value in units of its uncertainty. With a mean of 170.2 ± 0.5 GeV, where the statistical uncertainty is corrected for the effect of the resampling according to Eq. 5.37, the measured top quark mass is in excellent agreement with the nominal value of 170 GeV. Each of the 1,000 pseudo-experiments yields a top mass within a window of 10 GeV around the central mass, no outliers are

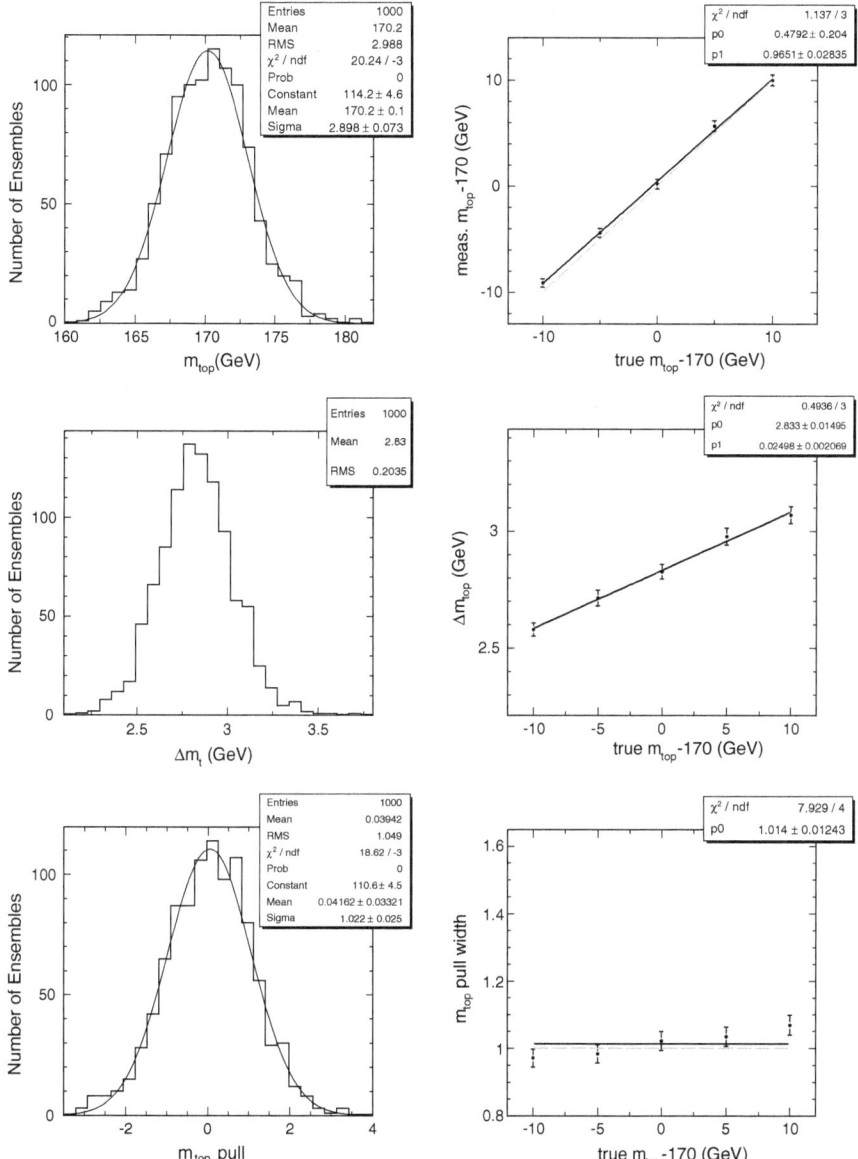

Fig. 6.8 Measurement of the top quark mass, its uncertainty, and its pull; *left* for the central mass point of 170 GeV; *right* as a function of the generated top masses. The two leptons are assumed to be perfectly measured, and the integration over the muon transverse momentum is not carried out. As the ensembles are purely built of $t\bar{t}$ events, no background likelihood is included

observed. For 50 signal events the statistical uncertainty is about 3 GeV, and the width of the pull distribution is in perfect agreement with 1. Thus, the statistical uncertainty is estimated correctly and the assumptions made for integration are perfectly valid.

On the right side of Fig. 6.8, the results for the additional mass points are shown. The mean of the top mass is always in good agreement with the nominal value. The uncertainty increases with the true top mass as expected and the pull width is flat over the full range. No mass-dependent effects are observed.

The corresponding plots are shown in Fig. 6.9 for the case of pure electron + muon events. One of the generated leptons is assumed to be an electron and perfectly measured, the other one a muon smeared according to the muon transfer functions. Accordingly, the integration over the inverse muon momentum is carried out. Within statistical uncertainties, no deviations from the case of perfectly measured electrons are observed. The mean of the mass is about 200 MeV higher where the uncertainty on the fit to the calibration points is already 300–400 MeV. The uncertainty is about the same and the pull width agrees perfectly with 1. Thus, in the following parton-level studies, both leptons are assumed to be perfectly measured. The integration over the inverse muon momentum is not carried out as no systematic bias is expected and no additional information on the method can be obtained by the integration, while it requires more computing time.

6.2.4 Studies Including $(Z \rightarrow \tau\tau)$ jj and $(Z \rightarrow \tau\tau)$ bb Events

In the next step, background events are included. In the electron + muon channel, the dominant source of background is expected to be $(Z \rightarrow \tau\tau)$ jj events, where the Z boson decays into two τ leptons and each of the τ leptons, into two neutrinos and a charged lepton. Compared to the leptons in top quark events, these leptons are on average significantly less energetic.

To estimate the effect of the $(Z \rightarrow \tau\tau)$ jj likelihood on the measured mass, two different scenarios are considered. First, $(Z \rightarrow \tau\tau)$ jj events are added but the background likelihood is not taken into account, then the same tests are performed and the background likelihood is included. On the left side of Fig. 6.10, the top quark mass, its uncertainty, and its pull are measured as a function of the $(Z \rightarrow \tau\tau)$ jj fraction in the ensembles. On the right side, the dependency on the true top mass is depicted. 50 events are used in each of the 1,000 pseudo-experiments, and the fraction of $(Z \rightarrow \tau\tau)$ jj events is varied from 10 to 50% in steps of 10%[2]. To study the dependency on the top mass, the fraction of $(Z \rightarrow \tau\tau)$ jj is fixed to 30% corresponding to the expected total fraction of background events in the electron + muon channel, see Sect. 6.1.

---

[2] As throughout this analysis, the fractions given in the text have to be understood as mean values where Poisson fluctuations are allowed.

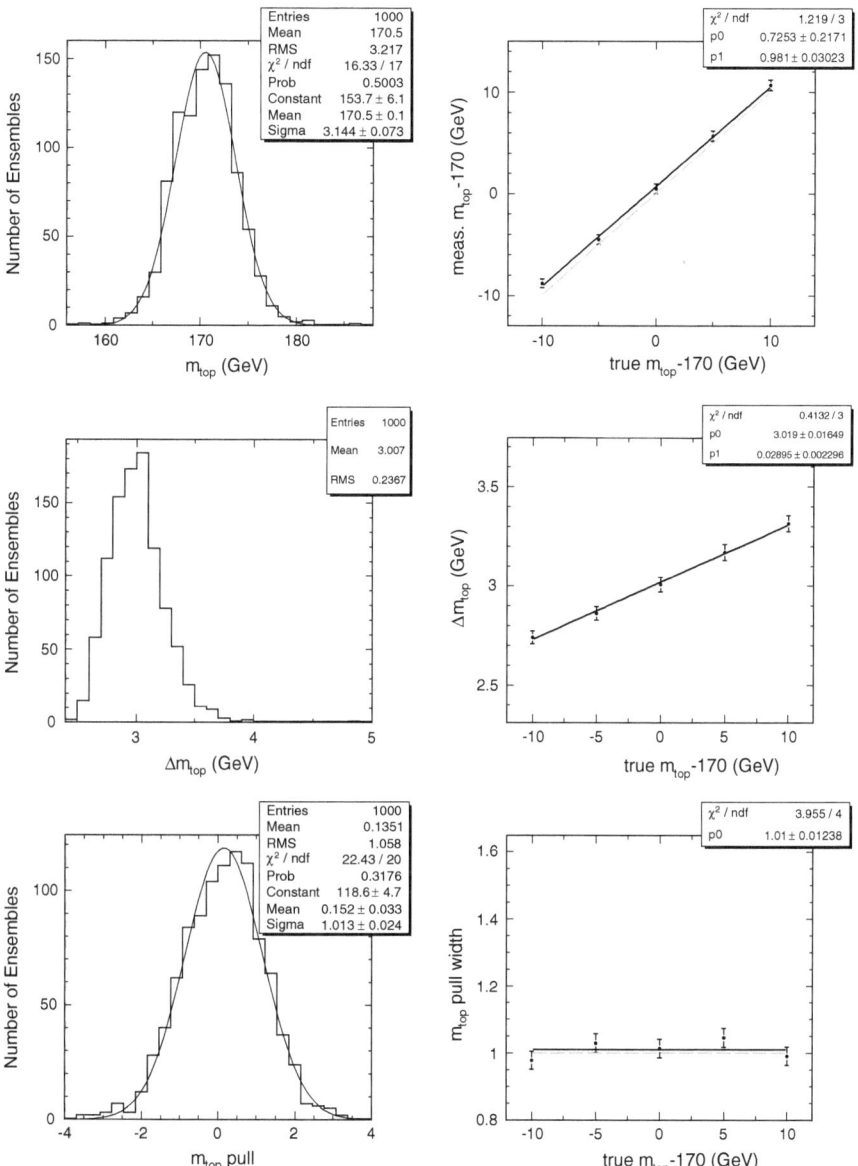

Fig. 6.9 Measurement of the top quark mass, its uncertainty, and its pull; *left* for the central mass point of 170 GeV; *right* as a function of generated top masses. One of the two leptons is smeared according to the muon transfer functions, and the additional integration is carried out. As the ensembles are purely built of $t\bar{t}$ events, no background likelihood is included

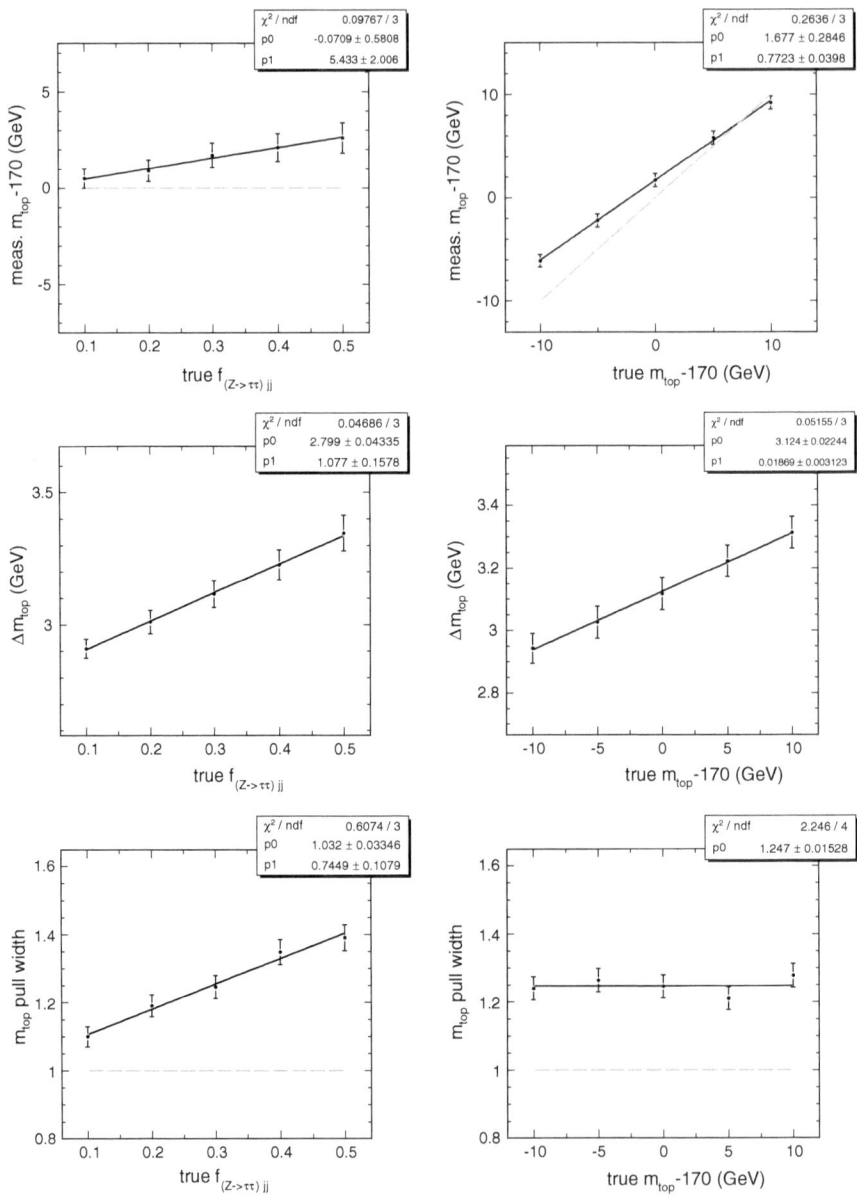

Fig. 6.10 Measurement of the top quark mass, its uncertainty, and its pull width; *left* as a function of the $(Z \to \tau\tau)\,jj$ fraction; *right* as a function of the true top mass. No background likelihood is included. For the mass dependency, the fraction of $(Z \to \tau\tau)\,jj$ events is fixed to 30%

When no background likelihood is included, the deviation of the measured top quark mass from its nominal value of 170 GeV increases with the fraction of $(Z \rightarrow \tau\tau)\, jj$ events. In the case of 50%, the measured mass is about 3 GeV off. The pull width increases to 40% since the assumptions made are not valid. For small top quark masses, the bias from the $(Z \rightarrow \tau\tau)\, jj$ events is largest.

In Fig. 6.11 however, the $(Z \rightarrow \tau\tau)\, jj$ likelihoods are included and the performance of the method is brought back to the level of signal-only events. Even in the case of 50% background, no bias is observed, and the pull width stays the same for all fractions of $(Z \rightarrow \tau\tau)\, jj$ events. The offset as a function of the generated top quark mass is about 400 MeV, and the pull width is in excellent agreement with 1.

To study the effect of jets from b quarks, additional $(Z \rightarrow \tau\tau)\, bb$ events are produced by smearing the partons with the b jet transfer functions. The left side of Fig. 6.12 shows the top quark mass, its uncertainty, and its pull as a function of $(Z \rightarrow \tau\tau)\, bb$ events, the right one, the dependency on the top quark mass. In the first case, the signal fraction is fixed to 70%, and the $(Z \rightarrow \tau\tau)\, bb$ fraction to the total number of background is varied from 10 to 50%. In the second, the absolute $(Z \rightarrow \tau\tau)\, bb$ fraction is fixed to 9% as expected for the electron + muon channel [8], and the signal fraction is chosen to be 70%. While no bias on the top quark mass is observed, the pull width declines slightly by about 3%. This can be explained by the fact that the generated partons are smeared according to the b jet transfer functions, whereas they are treated as light jets in the calculation of the background likelihoods. However, the effect is negligible, and the assumption of light-quark jets in the background likelihood is reasonable. Any possible bias on the top quark mass is accounted for by the final calibration curve used for the measurement as the fully simulated events include jets from b quarks.

6.2.5 Studies Including $(Z \rightarrow \tau\tau)\, jj$ and $WWjj$ Events

Additional contamination of the $t\bar{t}$ data sample comes from $WWjj$ events. The expected fraction in the electron + muon channel compared to $(Z \rightarrow \tau\tau)\, jj$ events is about a third. On the left side of Fig. 6.13, the dependency of the top quark mass, its uncertainty, and its pull width on the fraction of $WWjj$ events is depicted. The signal fraction is fixed to the expected signal fraction of 70%, see Sect. 6.1. The fraction of $WWjj$ events relative to the total number of background is varied from 10 to 50% in steps of 10%. Within statistical uncertainties, the measured top quark mass does not depend on the fraction of $WWjj$ events, while the pull width increases to 20%. This can be explained as the leptons in theses events really come from W decays and the reconstructed invariant top mass is about the same as in $t\bar{t}$ events. However, as no likelihood for this kind of events is included, the pull width increases indicating that the model used is not perfectly valid.

The right side of Fig. 6.13 shows the dependency of the measurement on the true top mass. On average, each of the 1,000 pseudo-experiments is composed by 35 signal, 12 $(Z \rightarrow \tau\tau)\, jj$ and 3 $WWjj$ events. A small dependency on the top quark mass

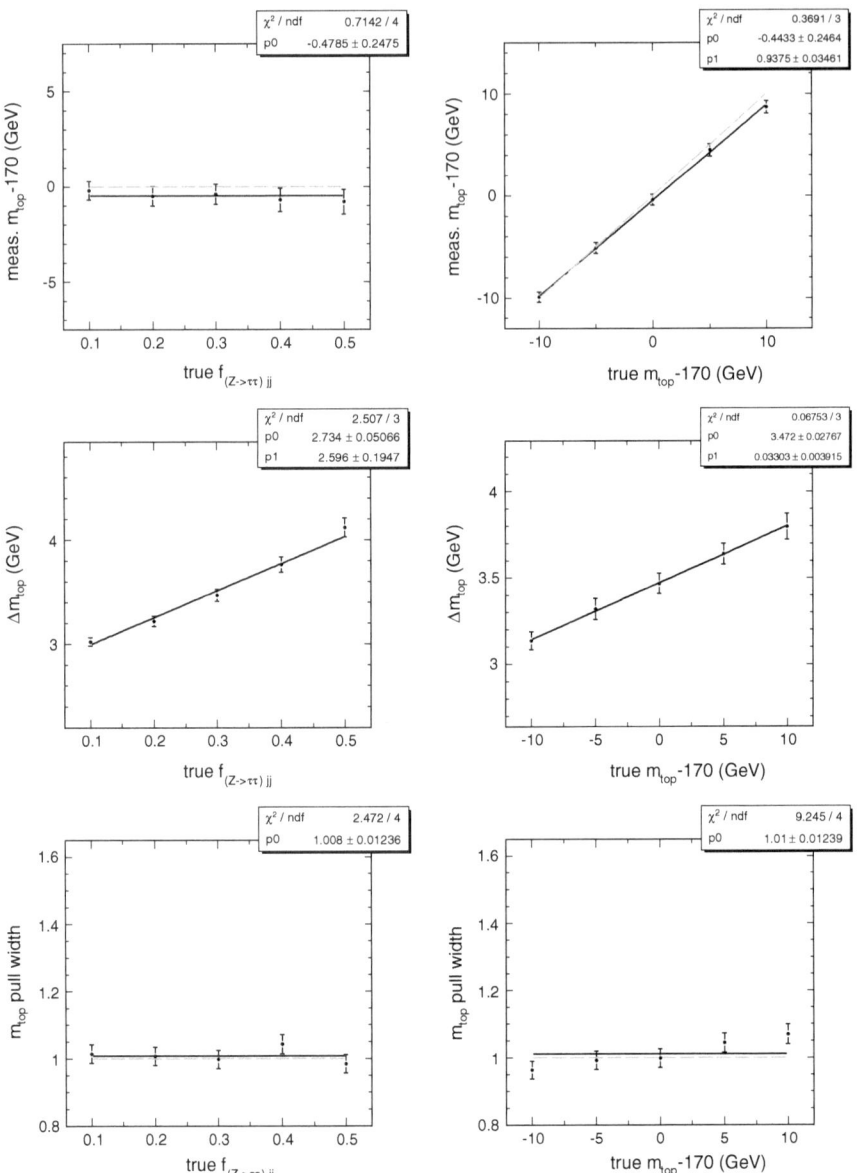

Fig. 6.11 Measurement of the top quark mass, its uncertainty, and its pull width; *left* as a function of the $(Z \rightarrow \tau\tau)\,jj$ fraction; *right* as a function of the true top mass. The $(Z \rightarrow \tau\tau)\,jj$ likelihood is included. For the mass dependency, the fraction of $(Z \rightarrow \tau\tau)\,jj$ events is fixed to 30%

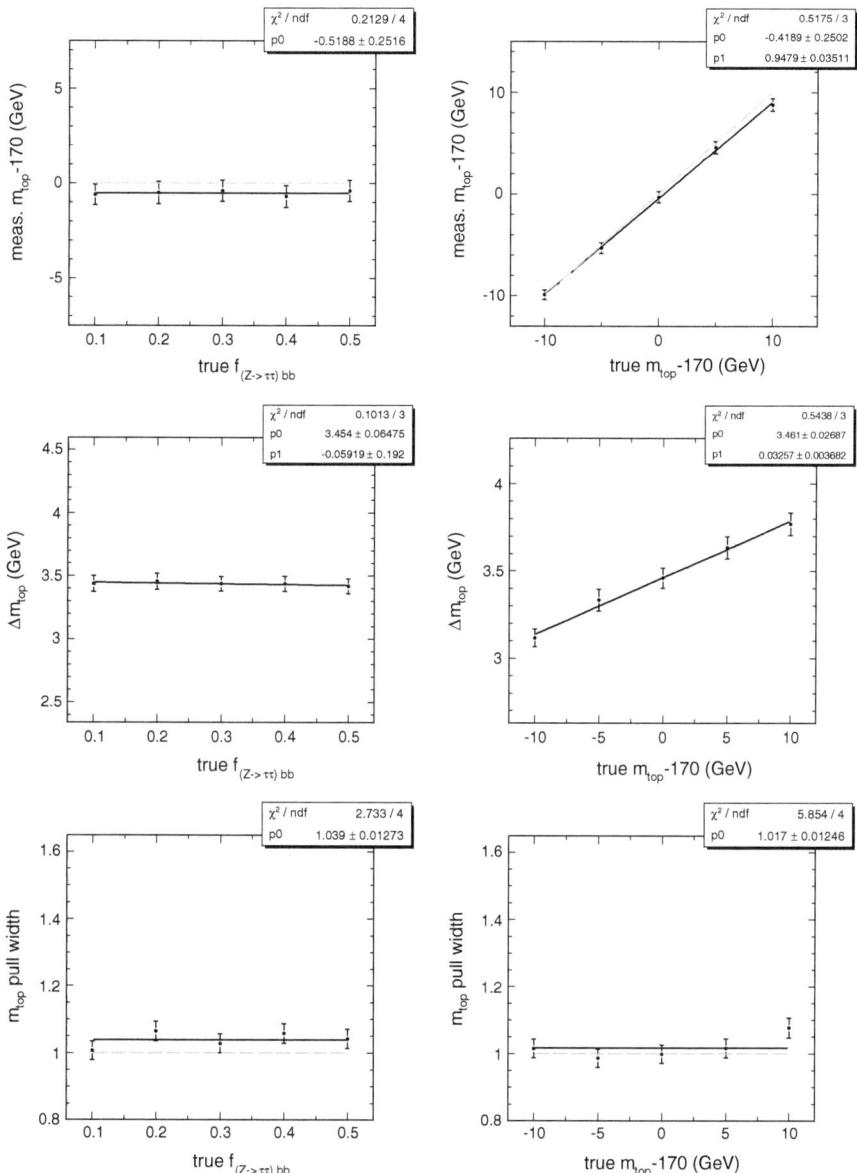

Fig. 6.12 Measurement of the top quark mass, its uncertainty, and its pull width; *left* as a function of the $(Z \to \tau\tau)$ bb fraction in background events; *right* as a function of the true top mass. For the mass dependency, the fraction of $(Z \to \tau\tau)$ bb events is fixed to 9%

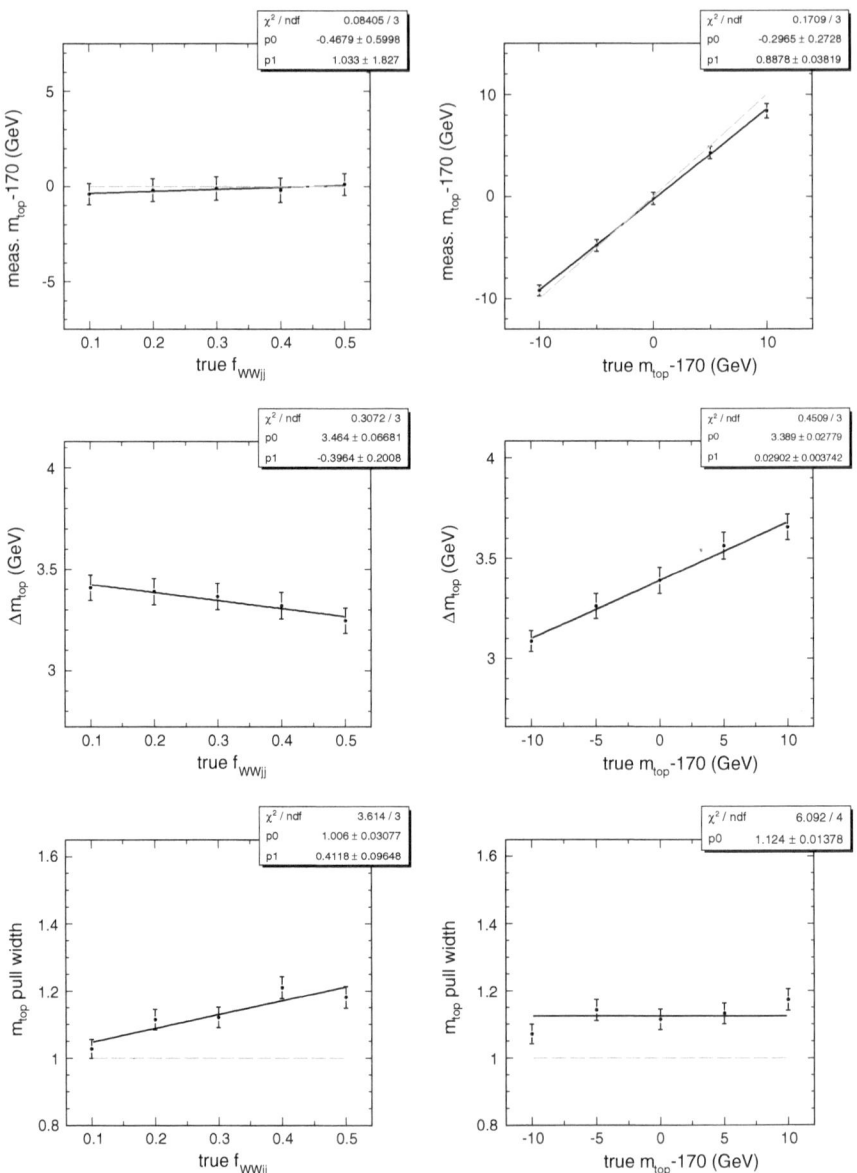

Fig. 6.13 Measurement of the top quark mass, its uncertainty, and its pull width; *left* as a function of the *WWjj* fraction in background events; *right* as a function of the true top mass. For the mass dependency, the fraction of *WWjj* is fixed to 6%

is observed, as the slope degrades to about 89%, while the offset for a true top quark mass of 170 GeV is still in good agreement with 0. This can be traced back to the fact that the invariant mass obtained by summing over jets and leptons in selected $WWjj$ events, is larger than the one in events with a true mass of 160 GeV, while it is lower than the one in events with a mass of 180 GeV. The pull width stays about the same within uncertainties for the different top quark masses.

As no significant degradation of the mass measurement is observed when the $WWjj$ likelihood is omitted from the event likelihood, this contribution is dropped to save computing time. Any effect due to this simplification is accounted for by the final calibration curve used for the measurement.

6.2.6 Measurement of the Signal Fraction

When background events are included, not only the top quark mass but also the fraction of signal events is measured. Strictly speaking, a parameter is measured for each process in the event likelihood, where all parameters sum up to 1 and the parameter ideally corresponds to the fraction of the process in the pseudo-experiment. The left side of Fig. 6.14 shows the measurement of the signal fraction as a function of the generated top mass, the right one as a function of the true signal fraction where the signal fraction is varied from 50 to 90% in steps of 10%. In the upper row, $t\bar{t}$ and $(Z \to \tau\tau)\,jj$ events are used, in the middle one, $t\bar{t}$, $(Z \to \tau\tau)jj$, and $(Z \to \tau\tau)\,bb$ events, in the lower one, $t\bar{t}$, $(Z \to \tau\tau)\,jj$, and $WWjj$ events. In case of $t\bar{t}$ and $(Z \to \tau\tau)\,jj$ events only, a perfect measurement of the signal fraction is possible as both processes are considered in the event likelihood. For a signal fraction of 70%, no bias is observed over the full range of top quark masses.

In the second scenario, the background is composed of 80% $(Z \to \tau\tau)\,jj$ events and 20% $(Z \to \tau\tau)\,bb$ events. Within statistical uncertainties, the measurement of the signal fraction works accurately. In the last setup, the background is composed of 70% $(Z \to \tau\tau)\,jj$ and 30% $WWjj$ events. As no $WWjj$ likelihood is included, the measured signal fraction is about 5% higher than the nominal value for all top masses. Again, the similarity of the $t\bar{t}$ and $WWjj$ events becomes visible. For 50% $t\bar{t}$, 35% $(Z \to \tau\tau)\,jj$, and 15% $WWjj$ events, the measured signal fraction is about 10% too large. The uncertainty on the signal fraction is accounted for in the systematic uncertainty by a systematic variation of the signal fraction.

6.3 Calibration of the Method

Since the Matrix Element method involves simplifications as the usage of a leading-order matrix element and the simplified treatment of the detector resolution via transfer functions, not every single aspect of the data can be accounted for and an adjustment of the raw result is mandatory. To calibrate the method, fully

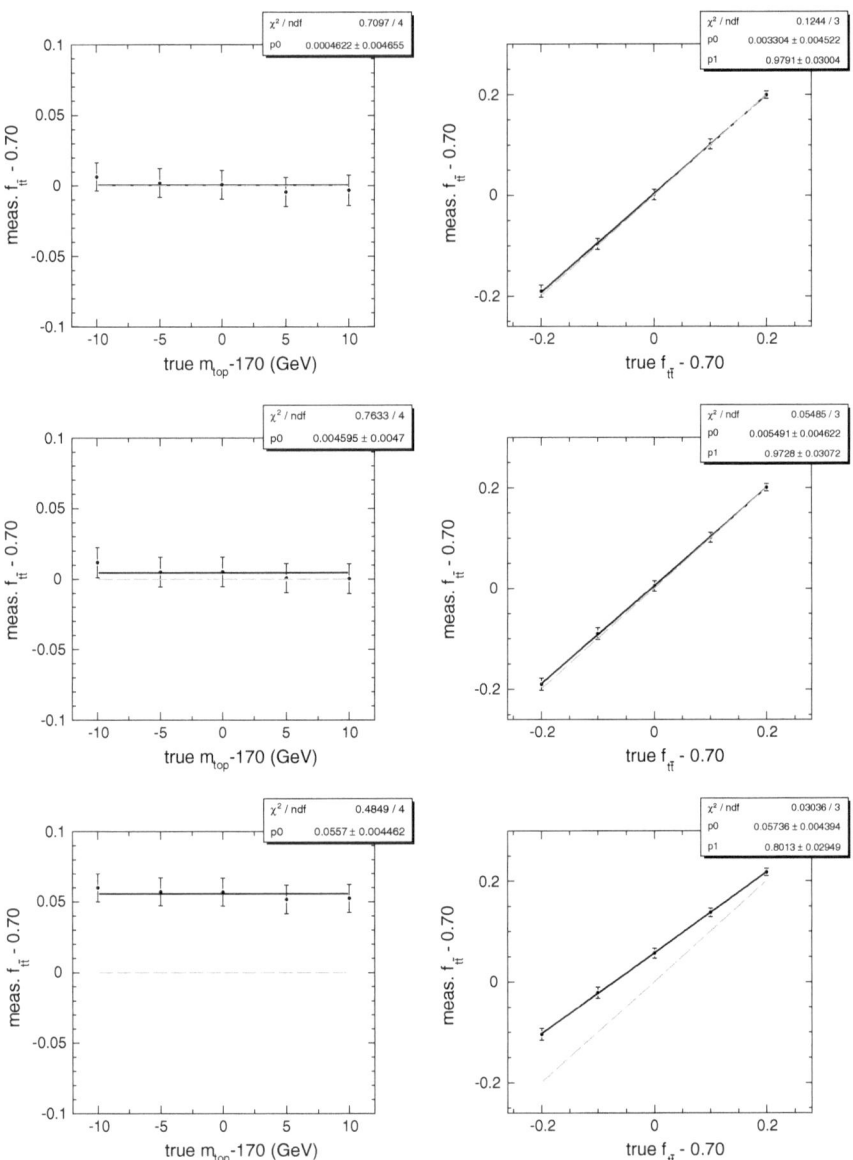

Fig. 6.14 Measurement of the signal fraction; *left* as a function of the generated top mass; *right* as a function of the true signal fraction. In the *upper row*, $t\bar{t}$ and $(Z \rightarrow \tau\tau) jj$ events are used, in the *middle one*, $t\bar{t}, (Z \rightarrow \tau\tau) jj$, and $(Z \rightarrow \tau\tau) bb$, and in the *lower one*, $t\bar{t}, (Z \rightarrow \tau\tau) jj$, and $WWjj$

simulated Monte Carlo samples of different top quark masses are generated and
the mass is measured in each sample. Thus, a calibration curve of the measured
top quark mass as a function of the generated one can be derived according to
Eq. 5.36.

6.3.1 Monte Carlo Samples

Unlike the Monte Carlo samples used in Sect. 6.2 for the step-by-step validation of
the method, the calibration curve is derived from events that are run through the
full detector simulation and that describe the selected data as discussed in
Sect. 6.1. For the measurement of the mass, signal samples are produced for six
top quark masses for Run IIa and five for Run IIb, i.e. 160, 165, 170, 175, 180,
185 GeV, and 150, 160, 170, 180, 190 GeV, respectively. The same mass points
could not be used for both run periods as they were not available.

As each $t\bar{t}$ signal sample consists of three different subsamples, $t\bar{t} + 0$lp, $t\bar{t} + 1$lp,
$t\bar{t} + 2$lp, the final $t\bar{t}$ samples are composed such that the right contribution from each
light-parton bin is reproduced and no weights need to be applied. To evaluate the
correct fractions, the matched cross sections are extracted from ALPGEN and the
selection efficiencies are measured in each light-parton bin. Thus, the fraction of
each subsample can be calculated as

$$f_j = \frac{\sigma_j \cdot \text{eff}_j}{\sum_{i=0}^{2} \sigma_i \cdot \text{eff}_i},$$

(6.1)

where eff_j is the selection efficiency of subsample j, and σ_j, the matched cross
section. The composition of the samples as well as the matched cross sections for
Run IIa and Run IIb are given in Tables 6.3 and 6.4. As the contribution from each
light-parton bin can vary with the top mass by up to 3.5% due to the matched cross
sections, it is checked that no effect on the top quark mass is observed when
varying the fractions within this range.

Table 6.3 Matched cross sections for the different light-parton bins from ALPGEN, as well as the
final contributions of each subsample j to the $t\bar{t}$ sample for Run IIa

Top Mass (GeV)	$t\bar{t} + 0$lp		$t\bar{t} + 1$lp		$t\bar{t} + 2$lp	
	σ (pb)	f (%)	σ (pb)	f (%)	σ (pb)	f (%)
160	5.15×10^{-1}	59.0	2.09×10^{-1}	26.7	1.06×10^{-1}	14.3
165	4.38×10^{-1}	58.0	1.84×10^{-1}	26.4	9.81×10^{-2}	15.6
170	3.76×10^{-1}	58.5	1.58×10^{-1}	26.4	8.36×10^{-2}	15.1
175	3.25×10^{-1}	59.4	1.35×10^{-1}	26.1	7.07×10^{-2}	14.5
180	2.83×10^{-1}	60.7	1.14×10^{-1}	26.5	5.54×10^{-2}	12.8
185	2.42×10^{-1}	59.9	1.01×10^{-1}	26.1	5.12×10^{-2}	14.0

The uncertainties of the different light-parton-bin fractions are about 2.5%

Table 6.4 Matched cross sections for the different light-parton bins from ALPGEN as well as the final contributions of each subsample to the $t\bar{t}$ sample for Run IIb

Top Mass (GeV)	$t\bar{t} + 0$lp		$t\bar{t} + 1$lp		$t\bar{t} + 2$lp	
	σ (pb)	f (%)	σ (pb)	f (%)	σ (pb)	f (%)
150	7.02×10^{-1}	57.4	3.02×10^{-1}	27.2	1.65×10^{-2}	15.4
160	5.15×10^{-1}	58.6	2.09×10^{-1}	27.0	1.06×10^{-1}	14.4
170	3.76×10^{-1}	59.4	1.58×10^{-1}	26.6	8.36×10^{-2}	14.0
180	2.83×10^{-1}	60.4	1.14×10^{-1}	26.3	5.54×10^{-2}	13.4
190	2.10×10^{-1}	61.0	8.80×10^{-2}	26.0	4.46×10^{-2}	13.0

The uncertainties of the different light-parton-bin fractions are about 2.5%

Table 6.5 Contribution from each sample to the pseudo-experiments for Run IIa and Run IIb

Period	$t\bar{t}$ (%)	$Z \to \tau\tau$ (%)	WW, WZ (%)	Data
Run IIa	83.0	13.4	3.6	39
Run IIb	84.4	11.6	4.1	68

The total number of events in each pseudo-experiment is chosen such that it reproduces the number of selected data events given in the last row

For the background simulation, $(Z \to \tau\tau) + \text{jets}$, $WW + \text{jets}$, and $WZ + \text{jets}$ events are included. Since the fraction of diboson events is small, these two processes are considered together. The effect of fake electrons and fake isolated muons is not accounted for in the pseudo-experiments as no proper samples are available. To allow for any possible bias due to this approximation a systematic uncertainty is assigned.

As in the case of the parton-level studies, the size of the event pool is limited by the computation time needed for the likelihood calculation. Additionally, for Run IIb it is restricted by the total number of selected events in case of the $(Z \to \tau\tau) + \text{jets}$ Monte Carlo sample. Nevertheless, no bias is observed from the resampling, and 30–40 independent pseudo-experiments can be built.

The number of events in each pseudo-experiment is chosen such that it corresponds to the observed data events. The fractions of each sample are derived from the data-to-Monte Carlo comparisons in Sect. 6.1, and listed in Table 6.5.

6.3.2 Normalization

The signal likelihoods are normalized following the procedure discussed in Sect. 5.6. Unlike for the parton-level events, the leading-order parton distribution functions CTEQ6L1 are used allowing for consistency with the PDFs in the event generation. To avoid any bias from the event selection, the following kinematic cuts are taken into account when calculating the likelihood normalization:

Fig. 6.15 Normalization for the signal likelihood as a function of m_{top} for the electron + muon channel in Run IIa and Run IIb. The difference between the two normalizations arise mainly from the different cuts on the transverse momentum of the leading jet. The effect of the different transfer functions is almost negligible

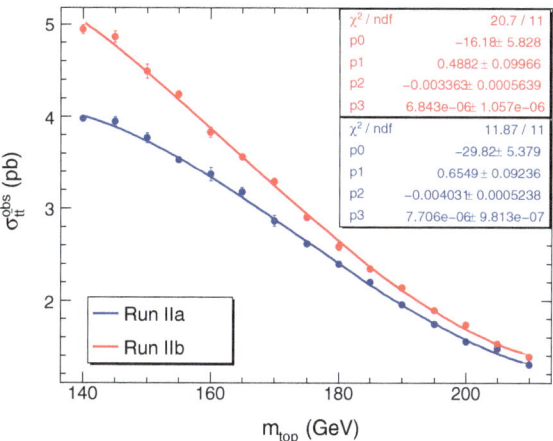

- $p_T^{j_{\text{lead}}} > 30(20)$ GeV for Run IIa (Run IIb)
- $p_T^{j_{\text{trail}}} > 20$ GeV
- $|\eta_j| < 2.5$
- $p_T^{\ell} > 15$ GeV
- $|\eta_e| < 1.1$ or $1.5 < |\eta_e| < 2.5$
- $|\eta_{\mu}| < 2.0$
- $p_T^{j_{\text{lead}}} + p_T^{j_{\text{trail}}} + p_T^{\ell_{\text{lead}}} > 115$ GeV
- $\Delta R(j, j') > 0.5$
- $\Delta R(\ell, j) > 0.5$

As the minimum transverse momentum of the leading jet is lowered to 20 GeV in Run IIb and the resolutions have changed between both run periods, two different normalization curves are derived. The fit and the parameters of the 3rd-order polynomial are given in Fig. 6.15.

The relative background-to-signal scale is derived as discussed in Sect. 5.8. The signal fraction is fixed to 80% as measured in Sect. 6.1. To minimize statistical fluctuations, the ensemble size is again maximized with respect to the available statistics for $t\bar{t}$ and $(Z \rightarrow \tau\tau)$ + jets events. The background normalization is taken as the mean of the relative scale measured as a function of the generated top quark mass. For Run IIa this is 30, for Run IIb, 100.

6.3.3 Validation of the Integration over the Top Pair Transverse Momentum

The fully simulated Monte Carlo events are used to validate the integration over the top pair transverse momentum as ALPGEN includes initial- and final-state radiation and the top pair system is not exclusively produced at rest in the transverse plane.

The left side of Fig. 6.16 shows the measured mass, its uncertainty, and its pull width in the case when the integration over $p_T^{t\bar{t}}$ is fixed, the right side, when it is carried out. For each, ensembles of 39 signal events are used. While the right mass curve is in perfect agreement with the expectation, an offset of about 2 GeV is observed for the left curve. As the additional momentum of the top pair system is not taken into account, the reconstructed mass is too high. In addition, the pull width is about 20% larger since the assumption of zero top pair transverse momentum is highly violated in events with more than two jets, see Fig. 5.10.

The remaining pull of about 15% is mainly caused by the assumption of equal direction for partons and jets. Due to the showering and the color reconnection with the beam remnant this is not true. Using only jet-parton-matched events where the distance between jets and partons in ΔR is required to be less than 0.5, the pull width can be reduced to about 5%.

6.3.4 Calibration for Run IIa and Run IIb

To calibrate the method for Run IIa and Run IIb, 1,000 pseudo-experiments are performed for each mass point with 39 or 68 simulated events, respectively. The fractions of the different samples given in Table 6.5 are determined from data-to-Monte Carlo comparisons. The contribution from fake electrons and fake isolated muons is not included as no proper samples are available. To account for statistical uncertainties, the sample fractions are allowed to fluctuate following Poisson statistics.

The left side of Fig. 6.17 depicts the result for Run IIa, the right one for Run IIb. As already seen in the parton-level studies in Sect. 6.2, the slope of the calibration curve degrades when adding background events. In Run IIa, no offset is observed, in Run IIb, 1%. This difference is mainly caused by the performance of the jet transfer functions. A good agreement is achieved when the true parton momenta are used. Already earlier studies [15] showed that the derivation and the performance of the transfer functions depends on several factors, as the top quark mass and the sample selection. As described in Sect. 5.3.1, due to available statistics, the same mass points could not be used. The uncertainty on the top mass is on the order of 4 GeV and slightly lower for Run IIb as more $t\bar{t}$ candidate events are available. The pull width for both Run IIa and Run IIb is about 15%. As already discussed in the last section, this is mainly due to the assumption of same direction for partons and jets.

The calibration curve for the signal fraction, $f_{t\bar{t}}$, and its dependency on the top quark mass are depicted in Fig. 6.18. For this study, the signal fraction is fixed to 83% in Run IIa and 84% in Run IIb. As shown in the parton-level tests, the discrepancy between the measured and the nominal signal fraction can be explained by the $WW +$ jets background, which is not accounted for in the event likelihood.

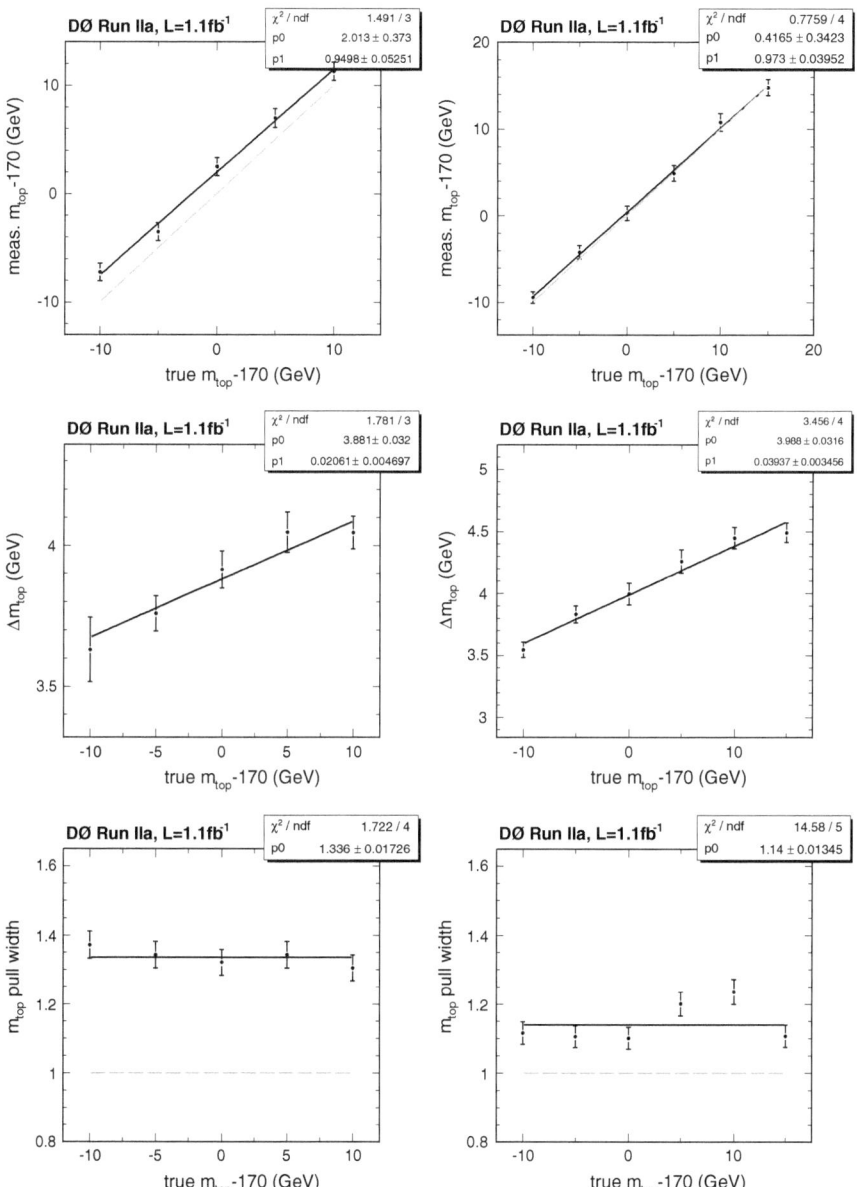

Fig. 6.16 Calibration curve for the top quark mass, its uncertainty, and its pull width using ensembles of pure signal events. *Left* the integration over the top pair transverse momentum is fixed. *Right* the integration is carried out. The solid lines show the fit to the points while the dashed ones show the perfect cases with no bias

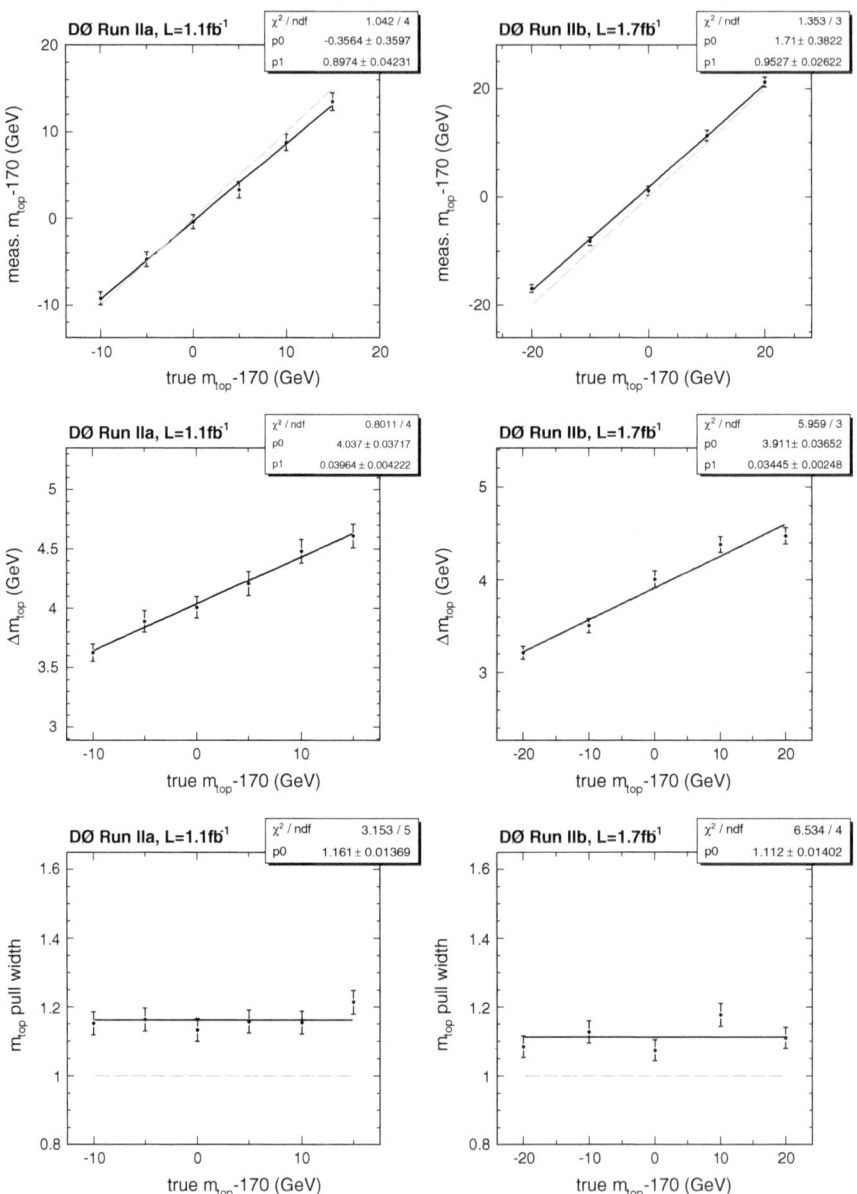

Fig. 6.17 Calibration curve for the top quark mass, its uncertainty, and its pull width; *left* for Run IIa; *right* for Run IIb. The *solid lines* shows the fit, while the *dashed lines* indicate the expectation for a bias free measurement

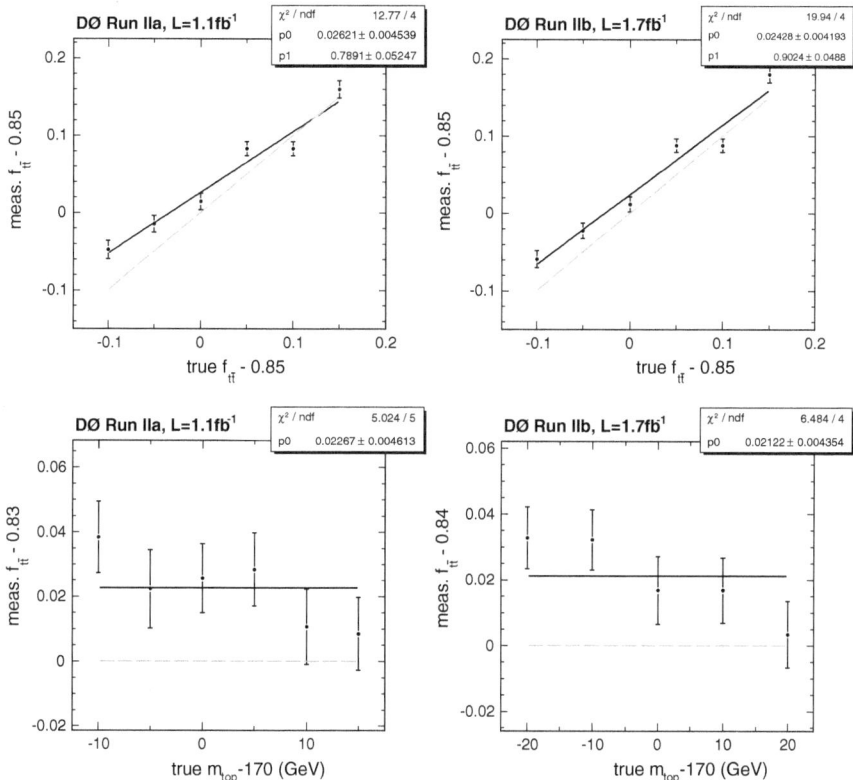

Fig. 6.18 Calibration curve for the signal fraction; *left* for Run IIa; *right* for Run IIb. The *upper row* shows the dependency on the true signal fraction, the *lower one* on the true top quark mass

Table 6.6 Parameters of the Run IIa and Run IIb calibration curves for the top quark mass and the signal fraction calculated according to Eq. 6.2

Period	$s_{m_{top}}$ (GeV)	$o_{m_{top}}$ (GeV)	$s_{f_{t\bar{t}}}$	$o_{f_{t\bar{t}}}$
Run IIa	0.90 ± 0.04	-0.36 ± 0.36	0.79 ± 0.05	0.026 ± 0.005
Run IIb	0.95 ± 0.03	1.71 ± 0.39	0.90 ± 0.05	0.025 ± 0.004

The parameters of the calibration curves for the top quark mass and the signal fraction calculated according to Eq. 5.36 for Run IIa and Run IIb are summarized in Table 6.6.

6.4 Measurement

In the present analysis, two different data sets are used to measure the top quark mass: the Run IIa and the Run IIb data samples collected between April 2002 and

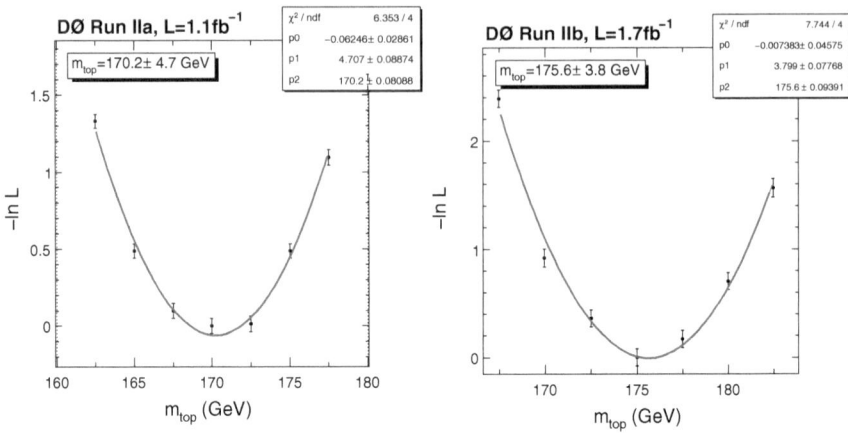

Fig. 6.19 Uncalibrated top quark mass measurement in the electron + muon channel; *left* for Run IIa; *right* for Run IIb

May 2008 at the DØ experiment. The first data set corresponds to an integrated luminosity of 1.1 fb^{-1}, the second, to an integrated luminosity of 1.7 fb^{-1}. Applying the selection cuts described in Sect. 6.1, 39 $t\bar{t}$ candidate events are available in Run IIa, and 68 data events in Run IIb. With a purity of about 78%, see Table 6.2, this corresponds to about 84 $t\bar{t}$ events.

The fits to the uncalibrated event likelihoods for the selected data samples are shown in Fig. 6.19. The statistical fluctuations are small and the likelihood is well described by a polynomial of the 2nd-order. In Run IIa, the mass is measured to be 170.2 ± 4.7 (stat.) GeV, in Run IIb, 175.6 ± 3.8 (stat.) GeV. To correct the raw measurements, the calibration curves shown in Fig. 6.17 are applied according to

$$m_{\text{top}}^{\text{cal}} = \left(m_{\text{top}}^{\text{raw}} - m_{\text{top}}^{\text{cent}} - o \right)/s + m_{\text{top}}^{\text{cent}} \qquad (6.2)$$

where the central mass $m_{\text{top}}^{\text{cent}}$ is 170 GeV. The slope, s, and the offset, o, of the calibration curves can be found in Table 6.6.

As discussed in Sect. 5.9, the statistical uncertainty is given by the 68% confidence region around the minimum of the likelihood distribution. To obtain the calibrated statistical uncertainty, the raw value is corrected by the slope of the calibration curve and the width of the pull distribution. The latter correction factor accounts for the underestimation of the statistical uncertainty evaluated in the ensemble-testing procedure.

Taking everything together, the top quark mass is measured to be

$$m_{\text{top}}^{\text{Run IIa}} = 170.6 \pm 6.1 \text{ (stat.)GeV} \qquad (6.3)$$

$$m_{\text{top}}^{\text{Run IIb}} = 174.1 \pm 4.4 \text{ (stat.)GeV.} \qquad (6.4)$$

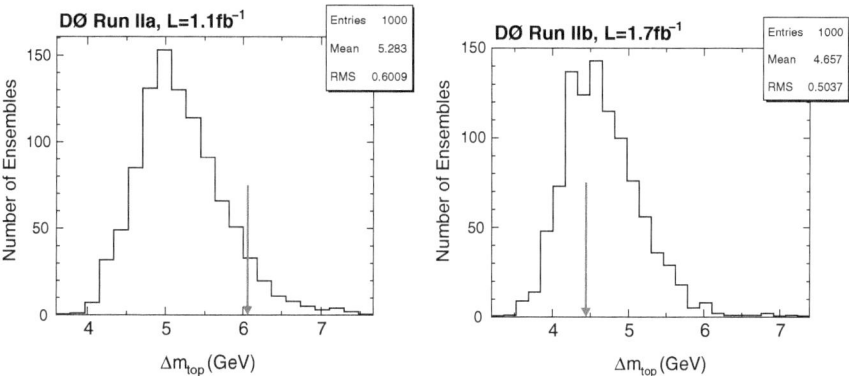

Fig. 6.20 Calibrated statistical uncertainties; *left* for Run IIa; *right* for Run IIb. The *arrows* indicate the statistical uncertainties measured in data

Table 6.7 Fitted top masses and signal fractions for the Run IIa and Run IIb data sets

Period	m_{top}^{raw} (GeV)	m_{top}^{cal} (GeV)	$f_{t\bar{t}}^{raw}$ (%)	$f_{t\bar{t}}^{cal}$ (%)
Run IIa	170.2 ± 4.7 (stat.)	170.6 ± 6.1 (stat.)	78.4 ± 7.3 (stat.)	73.9 ± 9.2 (stat.)
Run IIb	175.6 ± 3.8 (stat.)	174.1 ± 4.4 (stat.)	86.7 ± 4.9 (stat.)	84.2 ± 6.8 (stat.)

The values in the left columns correspond to the uncalibrated results, the ones in the right columns, to the calibrated results

The distributions of the calibrated statistical uncertainties from the ensemble tests for a generated top quark mass of 170 GeV are shown in Fig. 6.20. The red arrows indicate the statistical uncertainties found in the data samples. In Run IIa the uncertainty measured in data is about 1 GeV larger than the mean of the Monte Carlo distribution but well within expectations, in Run IIb, both values are about the same.

Besides the top quark masses, the signal fractions, $f_{t\bar{t}}$, are measured for both run periods and listed in Table 6.7. Correlations with the top quark mass are not taken into account when evaluating the statistical uncertainties. Applying the calibration curves given in Table 6.6, the signal fractions are measured to be

$$f_{t\bar{t}}^{Run\,IIa} = 73.9 \pm 9.2 \text{ (stat.)}\% \qquad (6.5)$$

$$f_{t\bar{t}}^{Run\,IIb} = 84.2 \pm 6.8 \text{ (stat.)}\%. \qquad (6.6)$$

Both measurements are in good agreement within their uncertainties with the purity of about 78% obtained from the data-to-Monte Carlo comparisons.

6.5 Systematic Uncertainties

Systematic uncertainties can affect the measured mass value in two different ways. First, the shape of the likelihood distribution can be biased directly. Secondly, the signal-to-background ratio of the selected data sample can be affected and thus the calibration curve shifted. Ideally, these two contributions should be treated coherently for each source of systematic uncertainty. However, in practice the second effect is much smaller than the first one for the most important systematic uncertainties. For simplicity, the sample composition systematics, i.e. the uncertainties on the signal-to-background ratio, are estimated using the combined systematic uncertainties from the measurements of the cross sections [8, 13]. Systematic uncertainties which are assumed to largely affect the mass likelihood distribution are treated individually, and the signal-to-background ratio is not modified.

Systematic uncertainties are evaluated in two different ways. For some uncertainties, the default ensembles are simply reweighted according to the expected uncertainty. Thus, a recalculation of the event likelihoods is not needed For other systematics however, this is not possible as they affect directly the input values of the event likelihoods, and a recalculation is required. In both cases, the difference between the measured top quark mass obtained when applying the default and the shifted calibration curves is taken as the systematic uncertainty.

Systematic uncertainties can be classified into three different categories: the modeling of the detector performance, the modeling of the physics processes for signal and background, and the uncertainties from the measurement methods. In the following, the three different sources of systematic uncertainties are discussed.

6.5.1 Detector Modeling

The largest systematic uncertainties in the measurement of the top quark mass arise from uncertainties on the detector response, mainly the determination of the jet energy scale. As simulated events are used to calibrate the method, only the relative uncertainty between data and simulation needs to be taken into account.

6.5.1.1 Overall Jet Energy Scale

This systematic uncertainty comes from the uncertainty on the ratio of the jet energy scales measured in data and simulated events, see Sects. 3.7.1 and 3.7.3. It is evaluated by shifting the jet energy scale corrections by ± 1 σ in the simulation where σ has been determined in the JSSR procedure [16]. Propagating the positive (negative) shift to the final result, the calibrated top mass is lower (higher) than the default one since the calibration curve is shifted to higher (lower) values, while the data measurement stays the same.

In the Run IIa measurement, this systematic uncertainty is evaluated to be $^{+1.2}_{-1.3}$GeV, while it is $^{+1.8}_{-1.5}$GeV in Run IIb. The shift between the two measurements can be traced back to the fact that the uncertainty on the jet energy scale has been inflated for Run IIb compared to Run IIa as not all detailed studies on the uncertainty are yet available for Run IIb. In a future measurement, this uncertainty will decrease for Run IIb.

6.5.1.2 Relative *b* Quark Jet Energy Scale

This error arises from the difference between the nominal inclusive response and the response of *b* jets. It accounts for the differences in the calorimeter response to electromagnetic and hadronic showers and the uncertainty on the electromagnetic and hadronic energy faction in *b* jets.

The relative uncertainty has been evaluated to be 1.8% [17], and the nominal response is multiplied with 0.982. As the evaluation requires a recalculation of the event likelihoods and the systematic variation is the same for Run IIa and Run IIb, it is only determined for Run IIa and the measured uncertainty of +1.4 GeV is assigned for both Run IIa and Run IIb.

6.5.1.3 Jet Resolution

The so called Jet Shifting, Smearing and Removal described in Sect. 3.7.3 applies an additional smearing to Monte Carlo jets to account for the different jet resolutions in data and Monte Carlo. To evaluate the effect of a residual uncorrected difference, the Monte Carlo generated events are smeared with a resolution varied by $\pm 1\sigma$ and a new set of calibration curves is derived.

The effect of the additional smearing is measured to be ± 0.6 GeV for Run IIa, and ± 0.7 GeV for Run IIb. Both values agree well within statistical uncertainties.

6.5.1.4 Jet Shifting

As already mentioned in Sect. 5.3.1, recent studies indicate that the jet energy shifting derived from $Z +$ jet events, see Sect. 3.7.3, over-corrects jets from *b* quarks. Since jets in $Z +$ jet events dominantly arise from gluons, they have a lower response and a wider shower than *b* jets.

The effect of this uncertainty is conservatively estimated by switching on the shifting correction and recalculating the event likelihoods. For Run IIa, the uncertainty is measured to be $+0.2$ GeV, for Run IIb, $+0.1$ GeV.

6.5.1.5 Lepton Resolution

As discussed in Sect. 3.5, an additional smearing is applied to leptons to account for the different resolutions in data and Monte Carlo. Since the uncertainty on the smearing is much larger for muons than for electrons [11, 18], the effect on the top quark mass is evaluated using smeared muons where the resolution is varied by $\pm 1\sigma$.

As this requires the event likelihoods to be recalculated, and as the size of the uncertainty is known to be small and to agree within statistical uncertainties between Run IIa and Run IIb, the value measured in Run IIa of $^{+0.3}_{-0.0}$ GeV is assigned for both run periods.

6.5.2 Physics Modeling

The uncertainties described in this section account for two different effects. First, many processes can lead to a signature similar to $t\bar{t}$ events in the detector, but not all of them can be taken into account in the simulation and the ensemble testing procedure. Secondly, the description of the processes that are accounted for may still be subject to uncertainties. Such additional uncertainties arise for example from the modeling of the parton distribution functions, the fragmentation of the b quarks, and the modeling of additional jets from initial- and final-state radiation.

6.5.2.1 Parton Distribution Functions

As the parton distribution functions yield the probability to find a parton of a certain flavor and momentum fraction inside the proton or antiproton, the kinematic distributions of the Monte Carlo events directly depend on the PDFs. The simulated events used to calibrate the measurement are based on the leading-order PDF set CTEQ6L1. However, systematic uncertainties are only provided for the next-to-leading order PDF set CTEQ6.1M. Therefore, the top quark mass is recomputed using a calibration based on the central CTEQ6.1M PDF set, and the difference between that value and the ones obtained with the systematic variations of the CTEQ6.1M PDF are added according to

$$\Delta m_{\text{top}}^{\text{PDF}+} = \sqrt{\sum_{i=1}^{20} \left(max\left(m_{\text{top},i}^+ - m_{\text{top}}^0, m_{\text{top},i}^- - m_{\text{top}}^0, 0 \right) \right)^2} \qquad (6.7)$$

$$\Delta m_{\text{top}}^{\text{PDF}-} = \sqrt{\sum_{i=1}^{20} \left(max\left(m_{\text{top}}^0 - m_{\text{top},i}^+, m_{\text{top}}^0 - m_{\text{top},i}^-, 0 \right) \right)^2} \qquad (6.8)$$

where m_{top}^0 denotes the default top mass, and $m_{\text{top}i}^+$ ($m_{\text{top},i}^-$) the one from the positive (negative) variation of the ith PDF eigenvector. This approach takes into account the sign of the variations propagated through to the observable, and the maximal positive and negative variations of the physical observable are considered separately. For more details see Bourilkov et al. [19].

Technically, each variation of a PDF eigenvector is implemented as an additional event weight which is taken into account when performing ensemble tests. For Run IIa the effect of the PDF uncertainties on the top quark mass is measured to be $^{+0.3}_{-0.0}$ GeV, and $^{+0.1}_{-0.2}$ GeV for Run IIb.

6.5.2.2 *b* Quark Fragmentation

Simulations based on different fragmentation and hadronization models predict different average energy fractions in reconstructed jets. This leads to an uncertainty on the relation between the jet and parton energies and thus the jet transfer functions and the measured top quark mass. Data from LEP and SLD on $Z \to b\bar{b}$ decays constrain b fragmentation models and yield a precise determination of the mean energy fraction of the weakly decaying bottom hadron [20]. To simulate top pair events and the corresponding distributions in top quark decays, different fragmentation models that are consistent with the Z data are used. To evaluate the effect of the fragmentation uncertainties on the top quark mass, the Monte Carlo simulated events are reweighted from the default PYTHIA b fragmentation to a Bowler scheme [21] that has been tuned to LEP or SLD data, respectively. The maximum difference between the results obtained from the reweighted samples to the unweighted sample is taken as the systematic uncertainty. For Run IIa, it is determined to be ±0.1 GeV, for Run IIb, ±0.3 GeV.

6.5.2.3 Signal Modeling

The main uncertainty in the modeling of the signal events comes from the uncertainty on the modeling of additional jets due to initial- and final-state radiation. As already discussed in Sect. 5.4, this has a large impact on the measured top mass. In case of initial-state radiation, the top pair system is no longer at rest in the transverse plane, while final-state radiation changes the momenta of the final-state b jets. In addition, ISR or FSR may lead to jets that can be misidentified as one of the $t\bar{t}$ decay products.

To evaluate the effect of the jet modeling on the measured top quark mass, the ratio of the numbers of events with exactly two jets (83%) and more than two jets (17%) is determined from data. The signal sample is then reweighted so that the ratio in the signal sample (75% events with exaclty two jets) matches the one in data. The difference between the measured mass based on the default calibration and the calibration from reweighted events is taken as the systematic uncertainty on the signal modeling. For Run IIa, the measured uncertainty is ±0.3 GeV, for Run IIb, ±0.4 GeV. As the top pair transverse momentum is accounted for in the calculation of the event likelihoods, this uncertainty is smaller than in other measurements of the top quark mass [22].

Table 6.8 Summary of systematic uncertainties

Uncertainty	Run IIa	Run IIb
JES up (GeV)	−1.3	−1.5
JES down (GeV)	+1.2	+1.8
b quark JES (GeV)	+1.4	+1.4*
Jet resolution up (GeV)	−0.6	−0.7
Jet resolution down (GeV)	+0.6	+0.7
ISSR shifting (GeV)	+0.2	+0.1
Muon smearing up (GeV)	−0.0	−0.0*
Muon smearing down (GeV)	+0.3	+0.3*
b quark fragmentation (GeV)	±0.1	±0.3
PDF uncertainty up (GeV)	−0.0	−0.2
PDF uncertainty down (GeV)	+0.3	+0.1
Signal modeling (GeV)	±0.3	±0.4
Background fraction up (GeV)	−0.0	−0.1
Background fraction down (GeV)	+0.2	+0.2
Calibration curve (GeV)	±0.4	±0.4
Total (GeV)	+2.1	+2.5
	−1.5	−1.8

* Taken from Run IIa

6.5.2.4 Background Fraction

Since the contribution of each source of events in the ensemble testing procedure is taken from the data-to-Monte Carlo comparisons discussed in Sect. 6.1, the uncertainties on the event yields shown in Table 6.2 may affect the derived calibration curve. Additionally, the contamination from the fake electron and fake isolated muon background is not taken into account when building ensembles.

To achieve a conservative estimate of this uncertainty with the Monte Carlo samples available, the number of background events is scaled up (down) by 1σ while scaling, at the same time, the number of signal events down (up) by 1σ. This allows to check the effect of a worst-case scenario in which the background contribution is underestimated while the signal contribution is overestimated. Two new calibration curves are derived taking into account the modified event yields and the difference between this result and the top quark mass based on the default calibration is assigned as the systematic uncertainty on the background fraction. For Run IIa, the effect is measured to be $^{+0.2}_{-0.0}$ GeV, for Run IIb, $^{+0.2}_{-0.1}$ GeV.

6.5.3 Uncertainties from the Measurement Method

The uncertainties described in this section account for any bias from the method itself. In the present analysis, this is only the uncertainty on the calibration curve.

6.5.3.1 Calibration Curve

To derive the calibration curve applied to the selected data sample, a polynomial of the first order is fitted to the calibration points of the top mass. Thus, the calibration curve depends on two parameters: the slope and the offset at the central value. Since both parameters can be only measured within the statistical uncertainty on the calibration points, this causes an uncertainty on the top mass.

To evaluate the effect from the uncertainty on the calibration curve the parameters are varied within their uncertainties and the top mass is measured based on the shifted calibration curves. The difference between this result and the default top mass is taken as the systematic uncertainty from the calibration curve. The measured uncertainty is ± 0.4 GeV for both Run IIa and Run IIb.

6.5.4 Summary of Systematic Uncertainties

Table 6.8 summarizes all systematic uncertainties on the top quark mass discussed in the last section.

The total systematic uncertainty is obtained by adding all positive and negative contributions separately in quadrature. The total systematic uncertainties for Run IIa and Run IIb are

$$\left(\Delta m_{\text{top}}\right)_{\text{syst}}^{\text{Run IIa}} = {}_{-1.5}^{+2.1} \text{ GeV} \tag{6.9}$$

$$\left(\Delta m_{\text{top}}\right)_{\text{syst}}^{\text{Run IIb}} = {}_{-1.8}^{+2.5} \text{ GeV}. \tag{6.10}$$

6.6 Combination of the Run IIa and Run IIb Mass Measurements

In the following, the combination of the two mass measurements presented in this analysis is described. The section starts with a short discussion of the so-called BLUE (Best Linear Unbiased Estimate) method which is used to combine the results. First, the principal idea of the method [23] is explained. Then, the extensions [24] needed in this context are given. The section ends with the discussion of the combination itself.

6.6.1 The Best Linear Unbiased Method

The easiest way to combine n different measurements of the same observable consists in weighting each result $y_i \pm \sigma_i$ by the inverse of its error squared σ_i^2 according to

$$\hat{y} = \left(\sum_i^n \frac{y_i}{\sigma_i^2}\right) \Big/ \left(\sum_i^n \frac{1}{\sigma_i^2}\right). \tag{6.11}$$

The corresponding error $\hat{\sigma}$ is then give n by

$$\frac{1}{\hat{\sigma}^2} = \sum_i^n \frac{1}{\sigma_i^2}. \tag{6.12}$$

Unfortunately, this applies only when the individual measurement errors are not correlated, and thus it cannot be used in this context where systematic correlations are present. The BLUE method allows however both correlated and uncorrelated errors. The method aims to construct a combined result $\hat{y}$ that fulfills the following requirements:

- The estimate $\hat{y}$ is a linear combination of the individual measurements with weights α_i to be determined:

$$\hat{y} = \sum_i^n \alpha_i \cdot y_i. \tag{6.13}$$

- $\hat{y}$ provides an unbiased estimate. Since all the results are assumed to be unbiased, the sum of all weights must be exactly one:

$$\sum_i^n \alpha_i = 1. \tag{6.14}$$

- The estimate $\hat{y}$ is the best amongst all estimates of the observable, i.e. the one of minimum variance. Following Eq. 6.13, the squared variance is given by

$$\sigma^2 = \vec{\alpha}' \mathbf{E} \vec{\alpha}, \tag{6.15}$$

where $\vec{\alpha}$ denotes the vector of the weights α_i, $\vec{\alpha}'$, its transpose, and $\mathbf{E}$, the error matrix of the measurements. The diagonal elements of $\mathbf{E}$ give the variances of the individual estimates, while the off-diagonal elements describe the correlations between pairs of estimates.

Given these constraints, the last step consists in finding the n values of α_i which minimize σ^2. This can be achieved using the method of Lagrangian multipliers. Thus, $\vec{\alpha}$ is given by

$$\vec{\alpha} = \mathbf{E}^{-1} \vec{u} / (\vec{u}' \mathbf{E}^{-1} \vec{u}), \tag{6.16}$$

where $\vec{u}$ is a vector whose n components are all unity, and $\mathbf{E}^{-1}$ is the inverse error matrix. This method is equivalent to minimizing the weighted sum of squares

$$S = \sum_i^n \sum_j^n (\hat{y} - y_{i,})(\hat{y} - y_{j,})(\mathbf{E}_{ij}^{-1}) \tag{6.17}$$

with respect to $\hat{y}$, where the weighted sum, S, measures the extent to which the individual measurements, y_i, are consistent with the combined result, $\hat{y}$. The final step in the procedure is now to use Eq. 6.17 to determine whether the individual estimates are self consistent or not. The weighted sum is expected to be distributed as χ^2 with $n - 1$ degrees of freedom and allows to judge how well the correlated results agree with each other.

Thus, the advantage of the BLUE method compared to a numerical approach is that it is simpler and more elegant: the weights α_i are determined a priori from matrix algebra and can then be used to compute both the combined result and its error. The BLUE method is an analytical solution to the problem of minimizing the χ^2 for the combination of measurements.

Besides the fact that the BLUE method also allows for combining different measured observables, an excellent feature is the simple breakdown of the error matrix into its individual components, such as those of statistical and systematic origins. Suppose that m independent sources of uncertainty have been identified, the total error matrix can be written as the sum of the different error matrices $\mathbf{E}_k$:

$$\mathbf{E} = \sum_{k}^{m} \mathbf{E}_k. \tag{6.18}$$

The individual contributions can now be propagated separately to the total error squared of the best estimate according to

$$\sigma^2 = \sum_{k}^{m} \vec{\alpha}' \mathbf{E}_k \vec{\alpha}. \tag{6.19}$$

The detailed breakdown of the error matrix is thus very simple to obtain with the BLUE method: the weights α_i are already available.

6.6.2 Combination

To combine both mass measurements, the uncertainty classes as defined by the Tevatron Electroweak Working Group (TEWWG) [25] are used. This includes a detailed breakdown of the various uncertainty sources and aims to combine sources of systematic uncertainties that share the same or similar origin. Some systematic uncertainties are split into multiple categories to accommodate specific types of correlations. For the combination, all uncertainties are symmetrized. In the following, the non-vanishing error classes relevant for this analysis as defined by the TEWWG are described:

- *aJES* aJES includes the part of the Jet Energy Scale uncertainty that arises from the difference in the electromagnetic-to-hadronic response between *b* jets and light jets. It corresponds to the uncertainty assigned in this thesis as b JES and is considered to be fully correlated between both measurements.

- *bJES* bJES is the part of the JES uncertainty which originates from uncertainties specific to the modeling of the b jets. It corresponds to the uncertainty on the b quark fragmentation and is considered to be fully correlated between both run periods.
- *dJES* This part of the JES uncertainty covers uncertainties from the limitations in the data samples used to derive the JES, i.e. the actual JES and the JSSR uncertainties. It is assumed to be fully correlated between both measurements.
- *Signal* This uncertainty includes uncertainties from the modeling of the $t\bar{t}$ Monte Carlo events, so the modeling of additional jets, ISR and FSR, and the parton distribution functions. It is taken to be fully correlated.
- *Bkg* This uncertainty summarizes the uncertainty from the background fraction as well as the jet and muon resolutions. It is assigned to be fully correlated between both measurements[3].
- *Fit* This uncertainty covers any uncertainty specific to a particular fit method. In the present analysis, the uncertainty on the calibration curves that are considered to be uncorrelated between both measurements belongs to this category.

To calculate the systematic uncertainties according to the TEWWG scheme, the values given in Table 6.8 are symmetrized and then added in quadrature in case of combined values. Applying the BLUE method with the statistical and systematic uncertainties as listed in Table 6.9, the combined value for the top quark mass determined using the Matrix Element method in the electron + muon final state is

Table 6.9 Top quark masses, statistical uncertainties, and systematic uncertainties according to the TEWWG classification scheme for Run IIa, Run IIb and the combination of both results

	Run IIa	Run IIb	Combination
m_{top} (GeV)	170.6	174.1	172.9
aJES (GeV)	1.40	1.40	1.40
bJES (GeV)	0.08	0.30	0.22
dJES (GeV)	1.27	1.65	1.52
Signal (GeV)	0.40	0.44	0.43
BG (GeV)	0.63	0.73	0.70
Fit (GeV)	0.40	0.40	0.30
Syst. (GeV)	2.07	2.38	2.25
Stat. (GeV)	6.10	4.40	3.57
Total (GeV)	6.44	5.00	4.22

---

[3] Even though it is the resolution uncertainty from the signal sample that mainly biases the top quark mass, it is not part of the signal uncertainty as in the combination of the world average the background uncertainty is fully correlated among all measurements in the same channel, while the signal uncertainty is fully correlated among all measurements.

$$m_{\text{top}} = 172.9 \pm 3.6 \text{ (stat.)} \pm 2.3 \text{ (syst.) GeV}$$
$$= 172.9 \pm 4.2 \text{ GeV.}$$

The combination yields a χ^2 of 0.2 for one degree of freedom which corresponds to a probability of 65%.

The measured top mass from Run IIa and Run IIb is in excellent agreement with the world average of 172.4 ± 0.7 (stat.) ± 1.0 (syst.) GeV [25].

References

1. Mangano ML et al (2003) ALPGEN, a generator for hard multiparton processes in hadronic collisions. JHEP 307:1
2. Sjöstrand T et al (2006) PYTHIA 6.4 physics and manual. JHEP 605:26
3. Hoche S et al (2006) Matching parton showers and matrix elements. arXiv:hep-ph/0602031
4. Lai HL et al (CTEQ collaboration) (2000) Global QCD analysis of parton structure of the nucleon: CTEQ5 parton distributions. Eur Phys J C 12:375
5. Kidionakis N et al (2003) Next-to-next-to-leading order soft-gluon corrections in top quark hadroproduction. Phys Rev D 68:114014
6. Hobbs JD et al (2006) Study of $p\bar{p} \to Z/\gamma \to ee$ and $\mu\mu$ event yields as a luminosity cross check. DØ note 5268
7. Martin B et al (2008) Final measurement of the $t\bar{t}$ production cross section at $\sqrt{s} = 1.96$ TeV in the ee final state using p17 data set. DØ note 5579
8. Arthaud M et al (2008) Final measurement of the $t\bar{t}$ production cross section at $\sqrt{s} = 1.96$ TeV in electron muon final states using p17 data set. DØ note 5580
9. Schellman H (2006) The longitudinal shape of the luminous region at DØ. DØ note 5142
10. Sharyy V. How to reweight MC according to the luminosity profile in data. https://plone4.fnal.gov/P1/D0Wiki/comp/caf/caaq/LumiReWeight
11. Calfayan P et al (2006) Muon identification certification for p17 data. DØ note 5157
12. Hays J et al (2006) The program package em_cert: version p18-br-20. DØ note 5070
13. Arthaud M et al (2008) $t\bar{t}$ event selection in the electron muon final states using p20 data set. DØ note 5720
14. Maltoni F, Stelzer T (2003) MadEvent: automatic event generation with MadGraph. JHEP 302:27
15. Schieferdecker P (2005) Measurement of the top quark mass at DØ Run II with the Matrix Element method in the lepton + jets final state. Dissertationsschrift, Ludwig-Maximilians-Universität München
16. Grivaz JF et al (2008) SSR for p17. DØ note 5609
17. Harel A (2008) Data over MC, b over light jet response corrections for Run IIa JES. DØ note 5654
18. Gris P (2007) Electron smearing studies with Run IIa data. DØ note 5400
19. Bourilkov D et al (2006) LHAPDF: PDF use from the Tevatron to the LHC. arXiv:hep-ph/0605240v2
20. The LEP collaborations, ALEPH, DELPHI, L3, OPAL, the SLD collaboration, the LEP Electroweak Working Group, and the SLD Electroweak, Heavy-Flavor Groups (2006) Precision electroweak measurements on the Z resonance. Phys Rep 427:257
21. Bowler MG (1981) $e^+ e^-$ production of heavy quarks in the String Model Z. Phys C 11:169
22. Fiedler F (2007) Precision measurements of the top quark mass. Habilitationsschrift, Ludwig-Maximilians-Universität München

23. Lyons L, Gibaut D (1988) How to combine correlated estimates of a single physical quantity. Nucl Instrum Meth A 270:110
24. Valassi A (2003) Combining correlated measurements of several different physical quantities. Nucl Instrum Meth A 500:391
25. The Tevatron Electroweak Working Group for the CDF and DØ collaborations (2008) Combination of CDF and DØ results on the mass of the top quark. arXiv:hep-ex/0703034

Chapter 7
Improved Mass Measurement

In the following chapter, a feasibility study for an improved measurement of the top quark mass in the dilepton channel will be discussed. As the largest systematic uncertainty on the top quark mass arises from the jet energy scale, a simultaneous fit of the mass and an additional scaling factor s_{bJES} on top of the nominal jet energy scale is performed. This allows for a stand-alone measurement of the b jet energy scale, and a significant reduction of the main systematic uncertainty. After a short introduction, the modifications are discussed in Sect. 7.2. To show the feasibility of the extended measurement, a large variety of parton-level tests have been carried out as described in Sect. 7.3.

7.1 Motivation

As seen in Sect. 6.5, the largest uncertainty on the measurement of the top quark mass arises from the uncertainty on the jet energy scale. With 1.9 GeV in Run IIa and 2.2 GeV in Run IIb, it is by far the dominant source of systematic uncertainties.

In the semileptonic decay channel [1], the impact of the jet energy scale is reduced by a simultaneous fit of the top quark mass, m_{top}, and the jet energy scale, s_{JES}. Here, s_{JES} is understood as an additional factor on top of the nominal jet energy scale described in Sect. 3.7.1; i.e. measuring s_{JES} to be 1.05 means that the jet energies are overestimated by 5%. If the directions of the light-quark jets from the W decay can be assumed to be perfectly measured, the measurement of s_{JES} is constrained by the nominal mass of the W boson. If the reconstructed W mass is in the order of the correct mass, the signal likelihood given in Eq. 5.26 is maximal. Analyzing many events, the additional scaling factor can be determined with high precision, and the main systematic uncertainty significantly reduced.

A. Grohsjean, *Measurement of the Top Quark Mass in the Dilepton Final State Using the Matrix Element Method*, Springer Theses, DOI: 10.1007/978-3-642-14070-9_7, © Springer-Verlag Berlin Heidelberg 2010

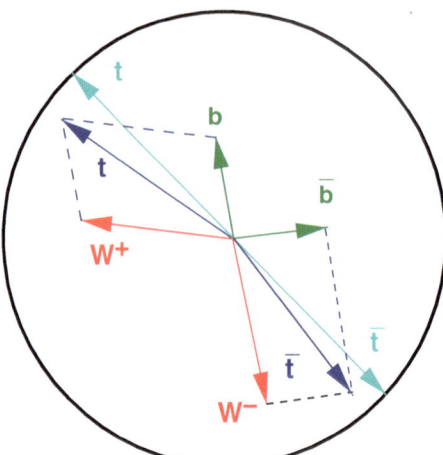

Fig. 7.1 Schematic drawing to illustrate the constraint that allows for the simultaneous measurement of the top quark mass and an additional b jet scaling. For the sake of simplicity, the top quarks (*light blue*) are assumed to be balanced in the transverse plane, and the W boson correctly reconstructed (*red*). If the b jet energies are underestimated by the b jet energy scale (*green*), the reconstructed top momenta (*dark blue*) are no longer balanced in the transverse plane, and the signal likelihood is low. As both jets are affected by the same s_{bJES}, only the correct scaling factor maximizes the signal likelihood

However, in case of an additional scaling factor for b jets, the constraint is much weaker. A qualitative picture of the constraint is given in Fig. 7.1. For the sake of simplicity, the top pair system is assumed to be at rest and the W boson correctly reconstructed. If the b jet energies are underestimated by the b jet energy scale, the reconstructed top momenta are no longer balanced in the transverse plane, and the signal likelihood is low. As both jets are affected by the same s_{bJES}, only the correct scaling factor maximizes the signal likelihood. The same constraint holds in case the top pair system is not at rest as long as the top pair transverse momentum is included in the likelihood calculation.

7.2 Modifications

As the Matrix Element method represents a generic approach to extract information about any set of parameters in the event likelihood, only minor modifications are needed to extend the method. In the following these modifications are discussed.

7.2.1 Event Likelihoods

For the simultaneous measurement of the top quark mass, m_{top}, and the b jet energy scale, s_{bJES}, the signal likelihood given in Eq. 5.26 is expanded to

$$L_{t\bar{t}}(x; m_{\text{top}}, s_{\text{bJES}}) = \frac{1}{\sigma_{t\bar{t}}^{\text{obs}}(m_{\text{top}}, s_{\text{bJES}})} \int_{\varepsilon_1, \varepsilon_2, \Phi_6, p_T^{t\bar{t}}} \sum_{\text{flavor,spin}} d\varepsilon_1 \, d\varepsilon_2 \, f_{\text{PDF}}(\varepsilon_1) f_{\text{PDF}}(\varepsilon_2)$$

$$\times \frac{(2\pi)^4 \left| \mathcal{M}_{t\bar{t}}(y; m_{\text{top}}) \right|^2}{\varepsilon_1 \, \varepsilon_2 \, s} \, d\Phi_6 \, W(x, y; s_{\text{bJES}}) \, dp_T^{t\bar{t}} \, W p_T^{t\bar{t}}.$$

$$(7.1)$$

Besides the observable cross section and the transfer functions, all parts of the method remain unchanged. The parameterization of the jet transfer functions given in Eq. 5.13 is replaced by

$$W_{\text{jet}}(\Delta E, s_{\text{bJES}}) = \frac{W_{\text{jet}}\left(\Delta E = \frac{E_{\text{jet}}^{\text{det}}}{s_{\text{bJES}}} - E_{\text{parton}}^{\text{ass}} \right)}{s_{\text{bJES}}}, \qquad (7.2)$$

where the factor s_{bJES} in the denominator ensures the correct normalization.

As the missing transverse energy is calculated according to Eq. 5.34 and used in the selection, the observable cross section also depends on s_{bJES}, and the normalization of the signal likelihood has to be changed accordingly, see Sect. 7.3.2.

Since more than 70% of the jets in background events arise from light quarks, both jets in the $(Z \to \tau\tau) jj$ background likelihood are treated as light jets, and the background likelihoods remain unchanged.

7.2.2 Normalization of the Jet Transfer Functions

The jet transfer functions can be normalized by integrating from zero to infinity or by integrating from the minimum jet momentum to infinity. In the first case, the cut on the transverse jet momentum has to be accounted for when calculating the observable cross section, in the second one, this is no longer needed if the event selection does not include a cut on the missing transverse energy. Since such a selection cut is applied in the analysis presented here, a dependency of the observable cross section remains even in the second case where the jet transfer functions are normalized from the minimum jet momentum to infinity.

Considering only parts of Eq. 7.1 that explicitly depend on s_{bJES}, the equivalence of both methods can be proved as follows:

$$
\int_{E_{\text{jet}}^{\text{det}}=20}^{\infty} dE_{\text{jet}}^{\text{det}} L_{t\bar{t}} = \frac{1}{\sigma_{\text{norm}}} \int_{E_{\text{jet}}^{\text{det}}=20}^{\infty} dE_{\text{jet}}^{\text{det}} \int_{E_{\text{parton}}^{\text{ass}}=0}^{\infty} d\sigma_{t\bar{t}}^{p\bar{p}}(E_{\text{parton}}^{\text{ass}}) \, W(E_{\text{jet}}^{\text{det}}, E_{\text{parton}}^{\text{ass}}; s_{\text{bJES}})
$$

$$
= \frac{1}{\sigma_{\text{norm}}} \int_{E_{\text{parton}}^{\text{ass}}=0}^{\infty} d\sigma_{t\bar{t}}^{p\bar{p}}(E_{\text{parton}}^{\text{ass}}) \int_{E_{\text{jet}}^{\text{det}}=20}^{\infty} dE_{\text{jet}}^{\text{det}} \, W(E_{\text{jet}}^{\text{det}}, E_{\text{parton}}^{\text{ass}}; s_{\text{bJES}})
$$

$$
= \frac{1}{\sigma_{\text{norm}}} \int_{E_{\text{parton}}^{\text{ass}}=0}^{\infty} d\sigma_{t\bar{t}}^{p\bar{p}}(E_{\text{parton}}^{\text{ass}})
$$

$$
= \frac{1}{\sigma_{\text{norm}}} \sigma_{t\bar{t}}^{p\bar{p}}.
$$

$$
\text{(7.3)}
$$

Since the likelihood $L_{t\bar{t}}$ must be normalized, the integral on the left-hand side is unity and the normalization factor

$$
\sigma_{\text{norm}} = \sigma_{t\bar{t}}^{p\bar{p}}. \tag{7.4}
$$

For consistency with a recent measurement of the top quark mass in the lepton + jets channel [2], the second method is used for the studies presented in this section. Assuming that only the cut on the minimum jet transverse momentum depends on s_{bJES}, the jet acceptance can be absorbed by the normalization of the jet transfer functions. However, as the missing transverse energy in the parton-level events is calculated from the negative sum of all detected particles, it also depends on s_{bJES}, and a two-dimensional normalization is still needed in the dilepton channel.

7.2.3 Likelihood Evaluation

The extraction of the top quark mass and the b jet energy scale is done as follows:

- First, the global maximum of the sample likelihood (5.3) is determined with respect to the signal fraction, $f_{t\bar{t}}$, by scanning the signal fraction in steps of 0.004 from 0 to 1 for all m_{top} and s_{bJES} hypotheses.
- With the fitted signal fraction, $f_{t\bar{t}}^{\text{best}}$, the sample likelihood is reduced to a two-dimensional function of m_{top} and s_{bJES}.
- For each m_{top} (s_{bJES}) hypothesis, the maximal likelihood value as a function of s_{bJES} (m_{top}) is determined.
- m_{top} (s_{bJES}) is then measured as the minimum of a polynomial fit of the second-order to the negative logarithm of the likelihood values. The fit range is chosen such that three m_{top} (s_{bJES}) hypotheses of either side around the minimum are included.
- The statistical uncertainties are given by the 68% confidence region around the measured values.

7.3 Parton-Level Studies

To validate the simultaneous measurement of the top quark mass, m_{top}, and the b jet energy scale, s_{bJES}, several tests are performed using parton-level events. In the following, these studies are discussed.

7.3.1 Monte Carlo Samples

In addition to the Monte Carlo samples described in Sect. 6.2.1, samples of different scaling factors s_{bJES} are produced for the default top quark mass of 170 GeV and the $(Z \rightarrow \tau\tau)\, bb$ sample by scaling the jet energies with 0.8, 0.9, 1.1, or 1.2 after the jet smearing. The missing transverse energy in the events is recalculated, and the kinematic selection, applied.

7.3.2 Normalization

To normalize the signal likelihoods, the cuts listed in Sect. 6.2.2 are taken into account; the cut on the jet transverse momentum is considered in the normalization of the jet transfer functions. As the missing transverse energy depends on the jet energy, see Eq. 5.34, the normalization of the signal likelihood depends on both m_{top} and s_{bJES}. The determination is done as follows:

For a given value of s_{bJES}, the observable cross section is calculated as a function of the top quark mass and fitted with a third-order polynomial, see Fig. 7.2. Each of the four parameters of the polynomials as a function of s_{bJES} are then fitted on their part with a quadratic function. The resulting two-dimensional normalization is depicted in Fig. 7.2.

The relative background-to-signal scale is rederived as described in Sect. 6.2.2 and the mean of the values obtained from all top mass samples is taken as the normalization scale for the background likelihoods.

7.3.3 Signal-Only Studies

In a first step, pseudo-experiments of pure signal events are performed, and the signal likelihood is taken as the event likelihood. The partonic jet energies are smeared according to the b jet transfer functions; the lepton momenta are assumed to be perfectly measured. Since the difference between smeared and unsmeared leptons does not affect the result when carrying out the integration over the inverse muon momentum and since no additional information can be obtained by such an integration, see Sect. 6.2.3, throughout this section, both leptons are assumed

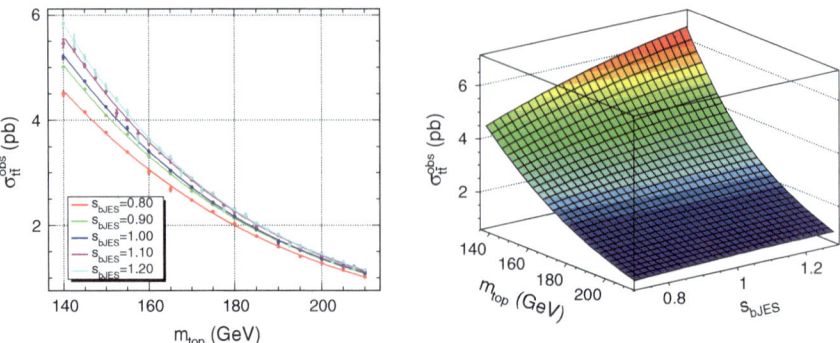

Fig. 7.2 *Left*: normalization curves for the signal likelihood as a function of the top quark mass for different values of s_{bJES}; *right*: the final two-dimensional parameterization

to be perfectly measured. In addition, the integration over the top pair transverse momentum is not carried out as the top pair system in the parton-level events generated by MADGRAPH [3] is produced at rest in the transverse plane.

1000 pseudo-experiments are performed for each of nine calibration points in the plane of (m_{top}(GeV), s_{bJES}), i.e. for (160, 1.0), (165, 1.0), (170, 1.0), (175, 1.0), (180, 1.0), (170, 0.8), (170, 0.9), (170, 1.1), and (170, 1.2). Each of the pseudo-experiments is built of $50t\bar{t}$ events.

On the left side of Fig. 7.3, the measured top quark mass, the statistical uncertainty, and the pull distribution for the central point of $m_{top} = 170$ GeV and $s_{bJES} = 1.0$ are shown, on the right side, the corresponding distributions for the b jet energy scale, s_{bJES}. With a mean of 170.4 ± 0.8 GeV, where the statistical uncertainty is corrected for the effect of the resampling according to Equation (5.37), the measured top quark mass is in excellent agreement with the nominal value. The same holds for s_{bJES} measured to be 1.00 ± 0.01.

However, it has to be noticed that only 970 out of the 1000 pseudo-experiments have a measured mass (b jet energy scale) within a window of 50 GeV (0.5) around the nominal value. This can be explained as each pseudo-experiment is composed of 50 events only, and the maximum of the sample likelihood as a function of s_{bJES} and m_{top} is not as distinct as it is for m_{top} only. Hence, it may happen that no clear maximum can be found and the measured values can be off in one pseudo-experiment.

The statistical uncertainty is about 6 GeV for the top quark mass, and 0.07 for the b jet energy scale. The statistical uncertainty is larger than in the one-dimensional fit as it now also includes the uncertainty on the jet energy scale. The widths of both pull distributions are in good agreement with 1.

The left side of Fig. 7.4 shows the measured mass, its uncertainty, and its pull width as a function of the generated top quark mass, the right one as a function of the generated b jet energy scale. Figure 7.5 depicts the corresponding plots for s_{bJES}. At all calibration points, the generated and the measured values are in excellent agreement. The uncertainty on m_{top} (s_{bJES}) is flat as a function of s_{bJES}

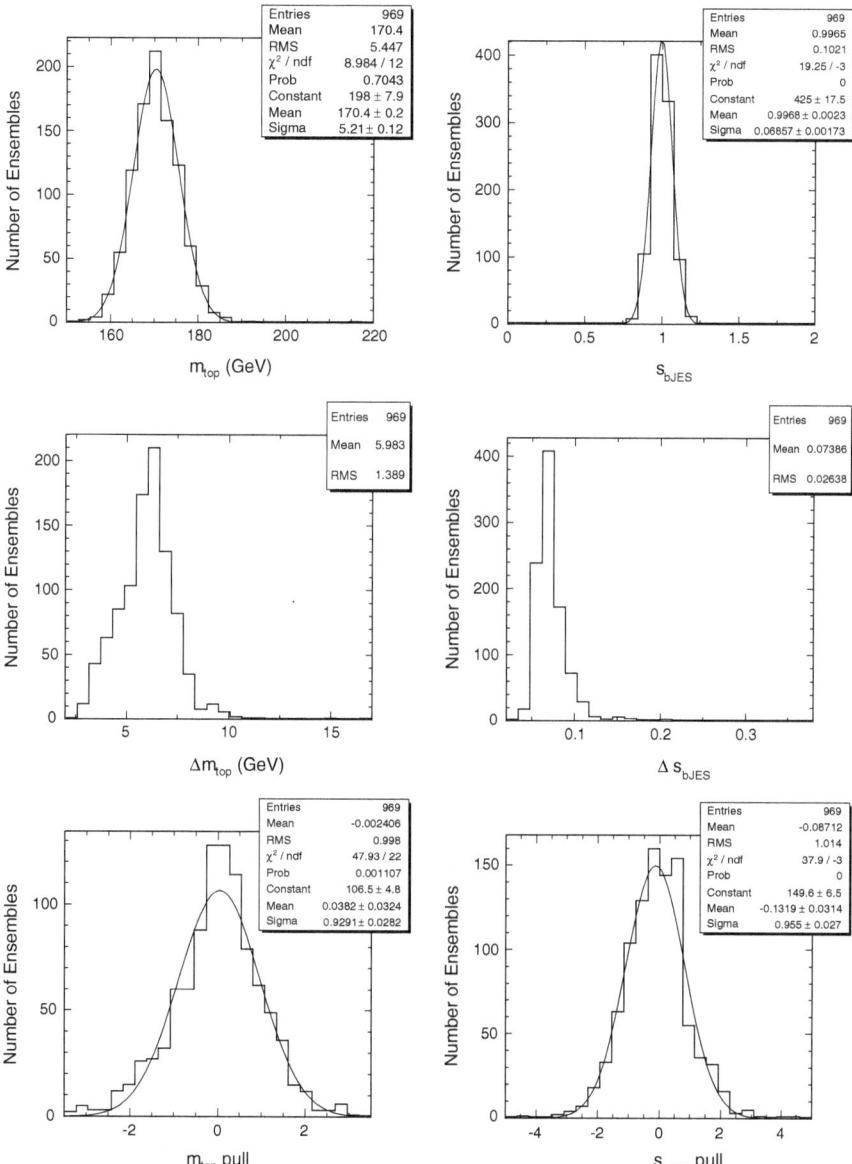

Fig. 7.3 *Left*: measurement of the top quark mass, its uncertainty, and its pull for the central sample of $m_{\text{top}} = 170$ GeV and $s_{\text{bJES}} = 1.0$; *right*: measurement of the b jet energy scale, its uncertainty, and its pull for the central sample. The two leptons are assumed to be perfectly measured, and the integration over the muon transverse momentum is not carried out. As the ensembles are purely built of $t\bar{t}$ events, no background likelihood is included

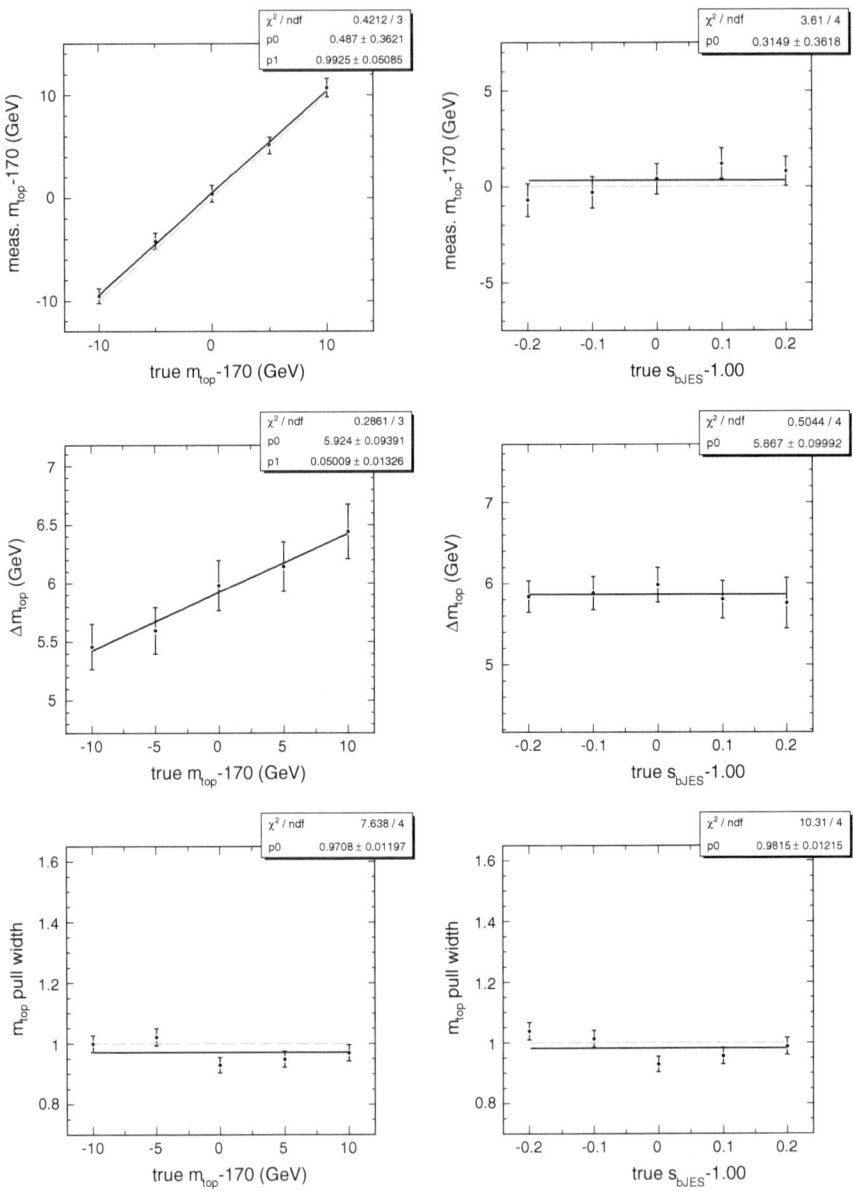

Fig. 7.4 Measurement of the top quark mass, its uncertainty, and its pull width; *left*: as a function of the top quark mass; *right*: as a function of the b jet energy scale

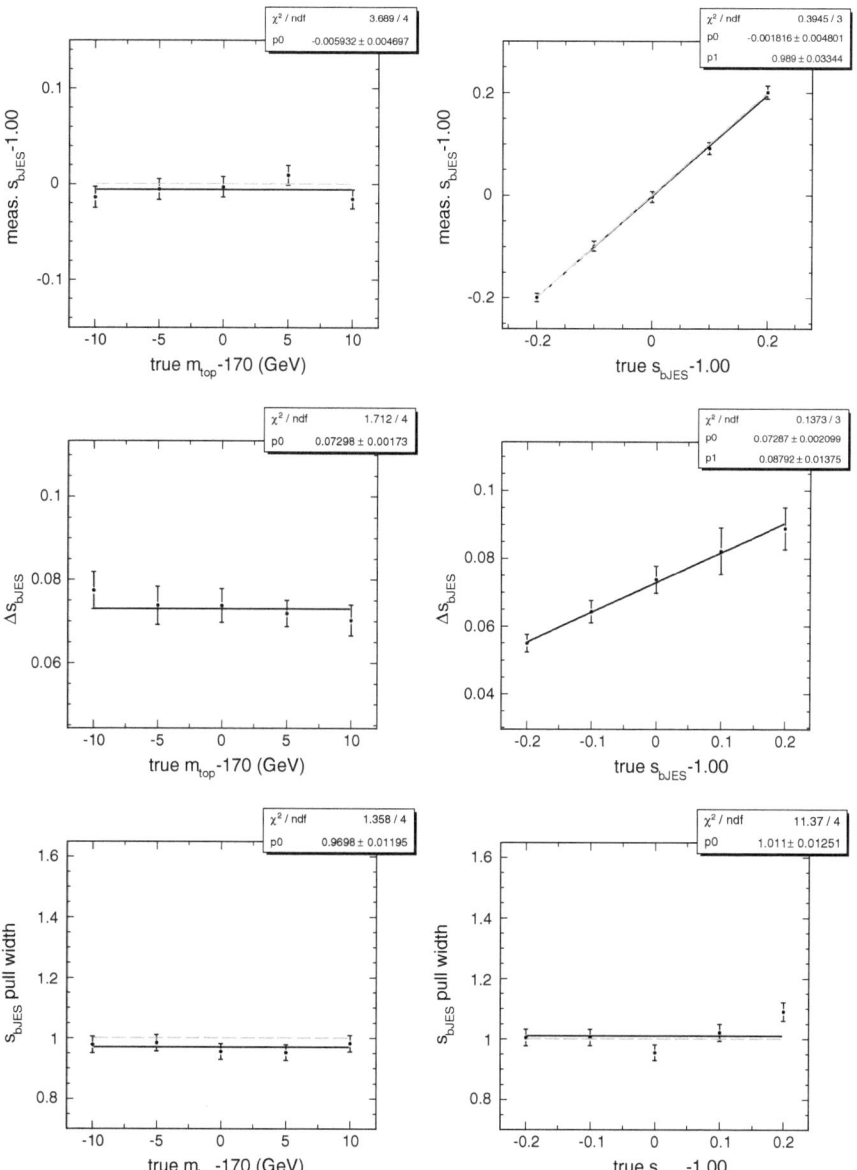

Fig. 7.5 Measurement of the *b* jet energy scale, its uncertainty, and its pull width; *left*: as a function of the top quark mass; *right*: as a function of the the *b* jet energy scale

(m_{top}), and increases with m_{top} (s_{bJES}) as expected. The pull width is always close to 1 within uncertainties. No mass or b jet energy scale dependent effects are observed when performing pseudo-experiments of pure signal events.

7.3.4 Studies Including $(Z \rightarrow \tau\tau)\, jj$ and $(Z \rightarrow \tau\tau)\, bb$ Events

In the next step, the dominant source of background is added, i.e. $(Z \rightarrow \tau\tau)\, jj$ events where the Z boson decays into an electron and a muon via two τ leptons. Accordingly, the $(Z \rightarrow \tau\tau)\, jj$ likelihood is included in the event likelihood.

On the left side of Fig. 7.6, the top quark mass, its uncertainty, and its pull width are measured as a function of the $(Z \rightarrow \tau\tau)\, jj$ fraction in the ensembles. On the right side, the same is done for the b jet energy scale. 50 events are used in each of the 1000 pseudo-experiments and the fraction of $(Z \rightarrow \tau\tau)\, jj$ events is varied from 10 to 50% in steps of 10%. Within statistical uncertainties, the measured values of m_{top} and s_{bJES} are almost flat when varying the fraction of $(Z \rightarrow \tau\tau)\, jj$ events. In the case of 50% background events, the top mass is measured to be $169.0 \pm 1.1\,$GeV, and the b jet energy scale to be 0.99 ± 0.02. Again, the stability of the result can be achieved by including the $(Z \rightarrow \tau\tau)\, jj$ background likelihood.

Figure 7.7 shows the calibration curve for m_{top} and s_{bJES}. The fraction of $(Z \rightarrow \tau\tau)\, jj$ is fixed to 30%. This corresponds to the total fraction of background events in the electron + muon channel, see Sect. 6.1. Both calibration curves are in excellent agreement with the expectation. The statistical uncertainty on m_{top} (s_{bJES}) increases with increasing m_{top} (s_{bJES}). The pull width of m_{top} is 0.99, and 1.03 for s_{bJES}. No deviation from the expectation is observed.

To study the effect of jets from b quarks, additional $(Z \rightarrow \tau\tau)\, bb$ events are included in the pseudo-experiments. The left side of Fig. 7.8 shows the top quark mass, its uncertainty, and its pull width as a function of the $(Z \rightarrow \tau\tau)\, bb$ fraction relative to the total number of background events, the right side, the same for the b jet energy scale. Within statistical uncertainties, no effect from the jet flavor can be observed when comparing Figs. 7.6 and 7.8. The top quark mass is only shifted additionally by about -200 MeV, while the b jet energy scale stays the same. As expected, no effect on the statistical uncertainty or the pull width is observed.

Figure 7.9 depicts the calibration curves for m_{top} and s_{bJES}. The $(Z \rightarrow \tau\tau)\, bb$ fraction is fixed to 9%, and the signal fraction is chosen to be 70%. No bias on the top quark mass or the b jet energy scale is observed, the pull width of m_{top} decreases by about 2%, while it stays the same for s_{bJES}. However, the effect is negligibly small, and the assumption of light-quark jets in the background likelihood is also reasonable for the simultaneous measurement of the top quark mass and the b jet energy scale.

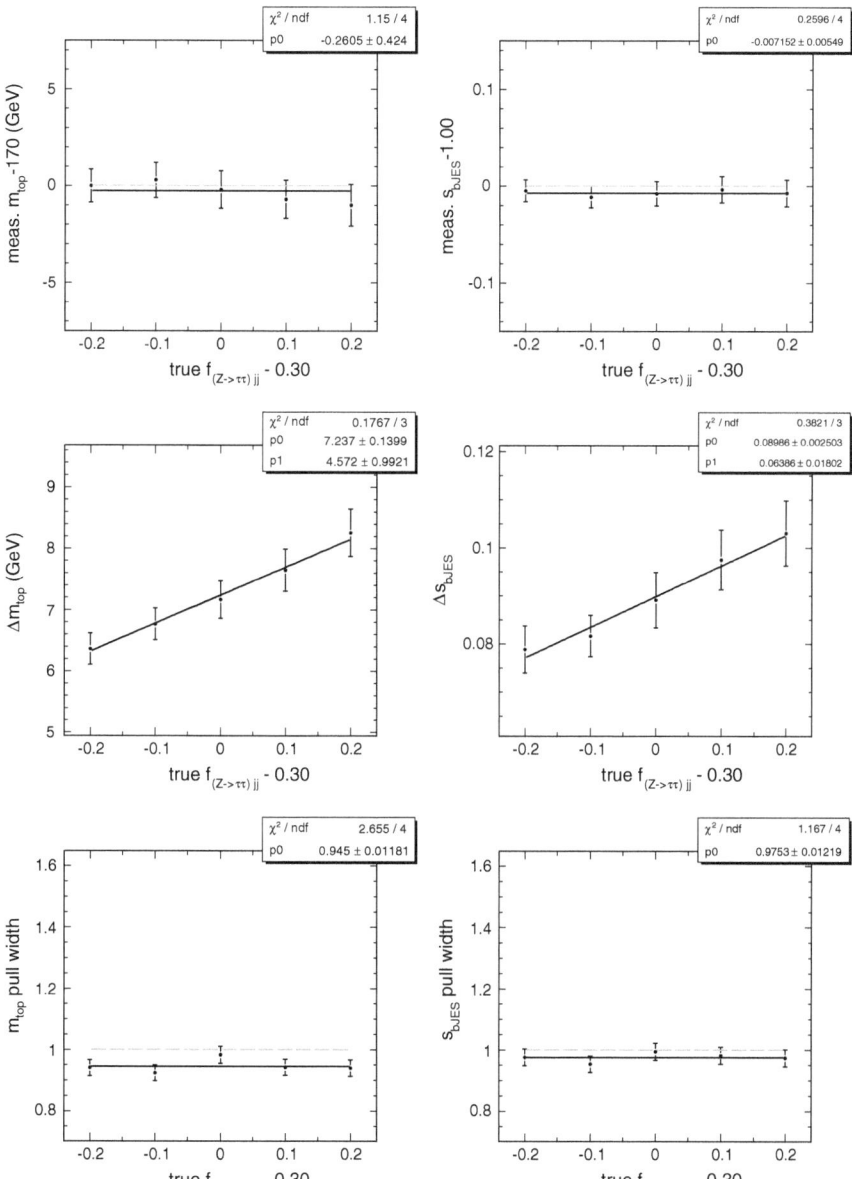

Fig. 7.6 *Left*: measurement of the top quark mass, its uncertainty, and its pull width as a function of the $(Z \to \tau\tau)jj$ fraction; *right*: measurement of the b jet energy scale, its uncertainty, and its pull width as a function of the $(Z \to \tau\tau)jj$ fraction

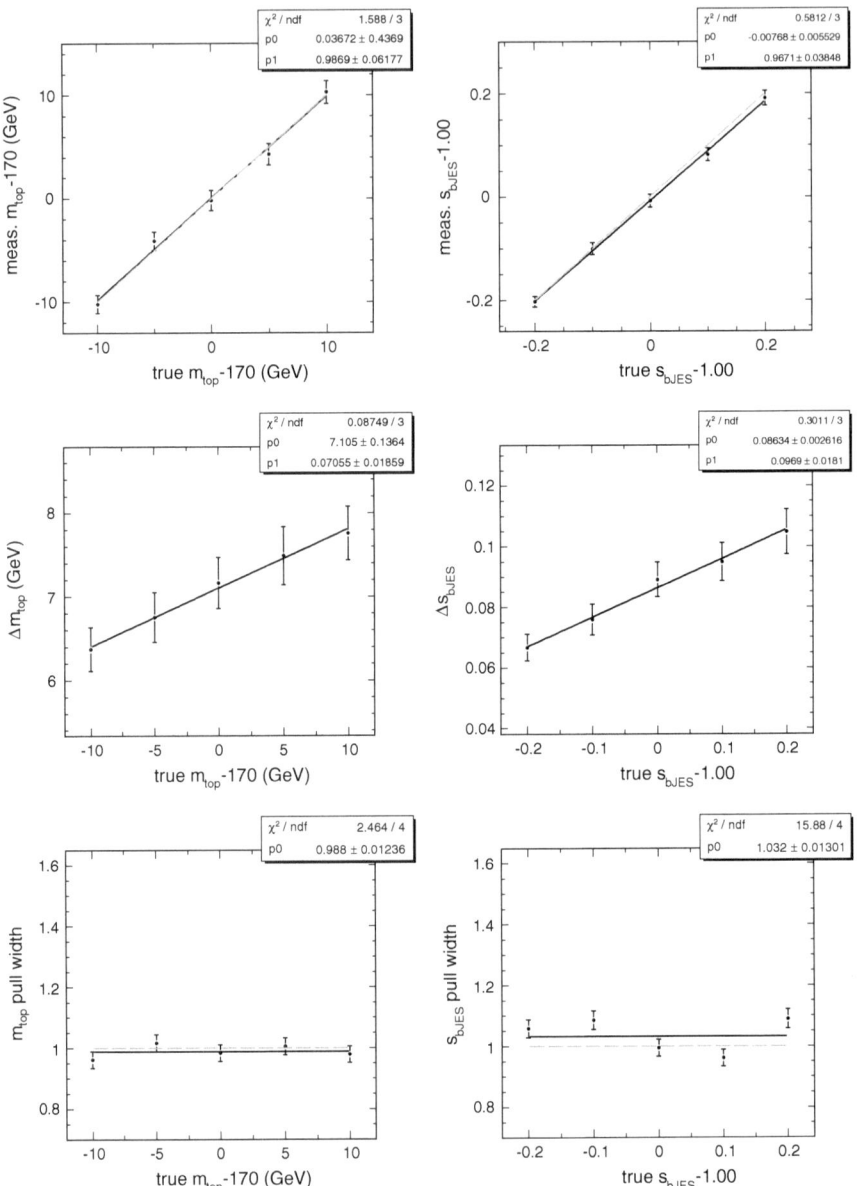

Fig. 7.7 *Left*: measurement of the top quark mass, its uncertainty, and its pull width as a function of the top quark mass; *right*: measurement of the b jet energy scale, its uncertainty, and its pull width as a function of the b jet energy scale. The fraction of $(Z \to \tau\tau)jj$ events is fixed to 30%

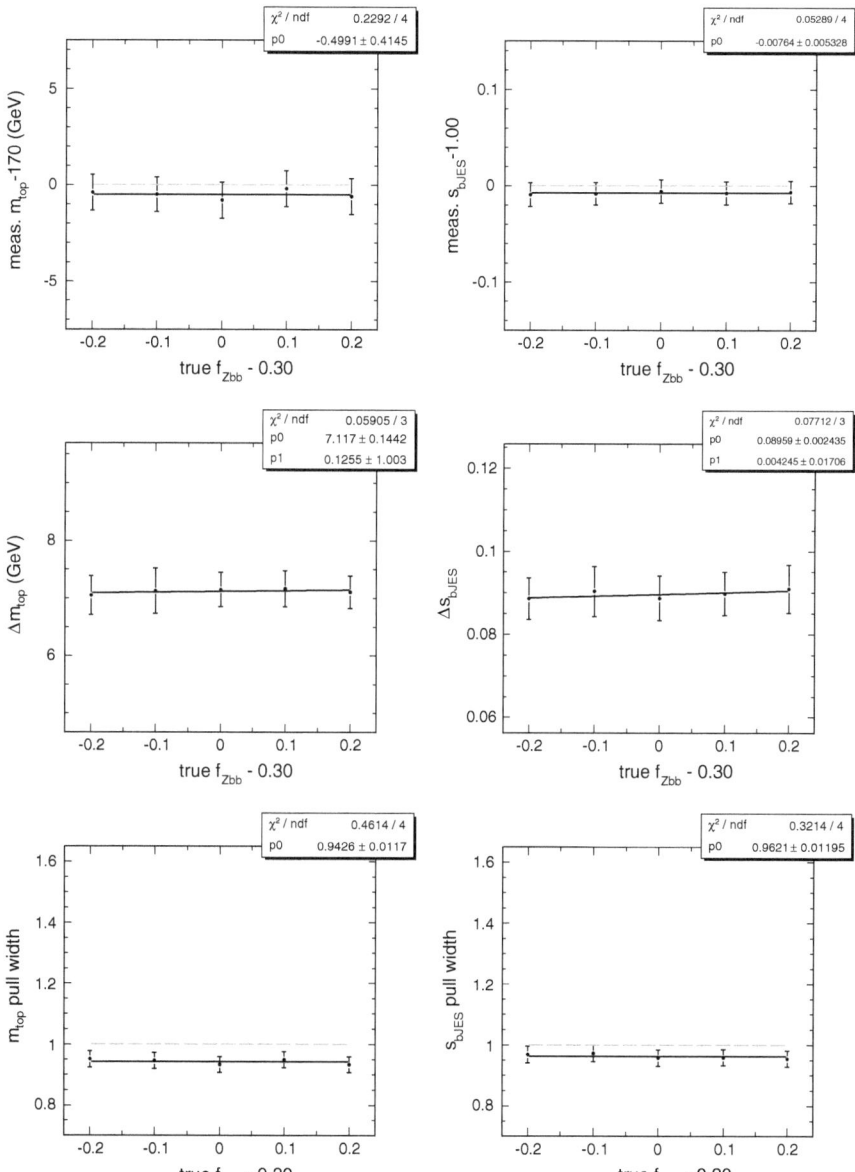

Fig. 7.8 *Left*: measurement of the top quark mass, its uncertainty, and its pull width as a function of the $(Z \to \tau\tau)\,bb$ fraction in background events; *right*: measurement of the b jet energy scale, its uncertainty, and its pull width as a function of the $(Z \to \tau\tau)\,bb$ fraction in background events. The signal fraction is fixed to 70%

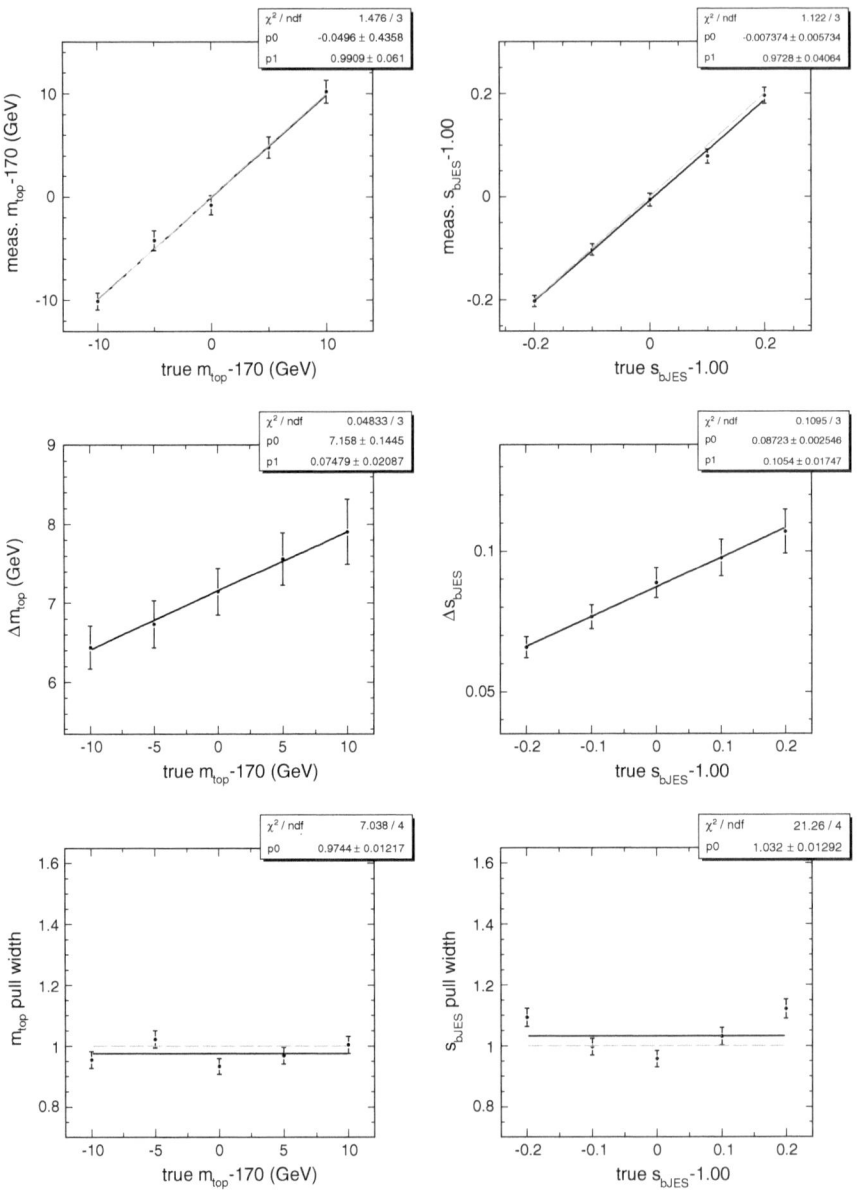

Fig. 7.9 *Left*: measurement of the top quark mass, its uncertainty, and its pull width as a function of the top quark mass; *right*: measurement of the b jet energy scale, its uncertainty, and its pull width as a function of the b jet energy scale. On average, 70% $t\bar{t}$, 21% $(Z \rightarrow \tau\tau)\,jj$ and 9% $(Z \rightarrow \tau\tau)\,bb$ events are used

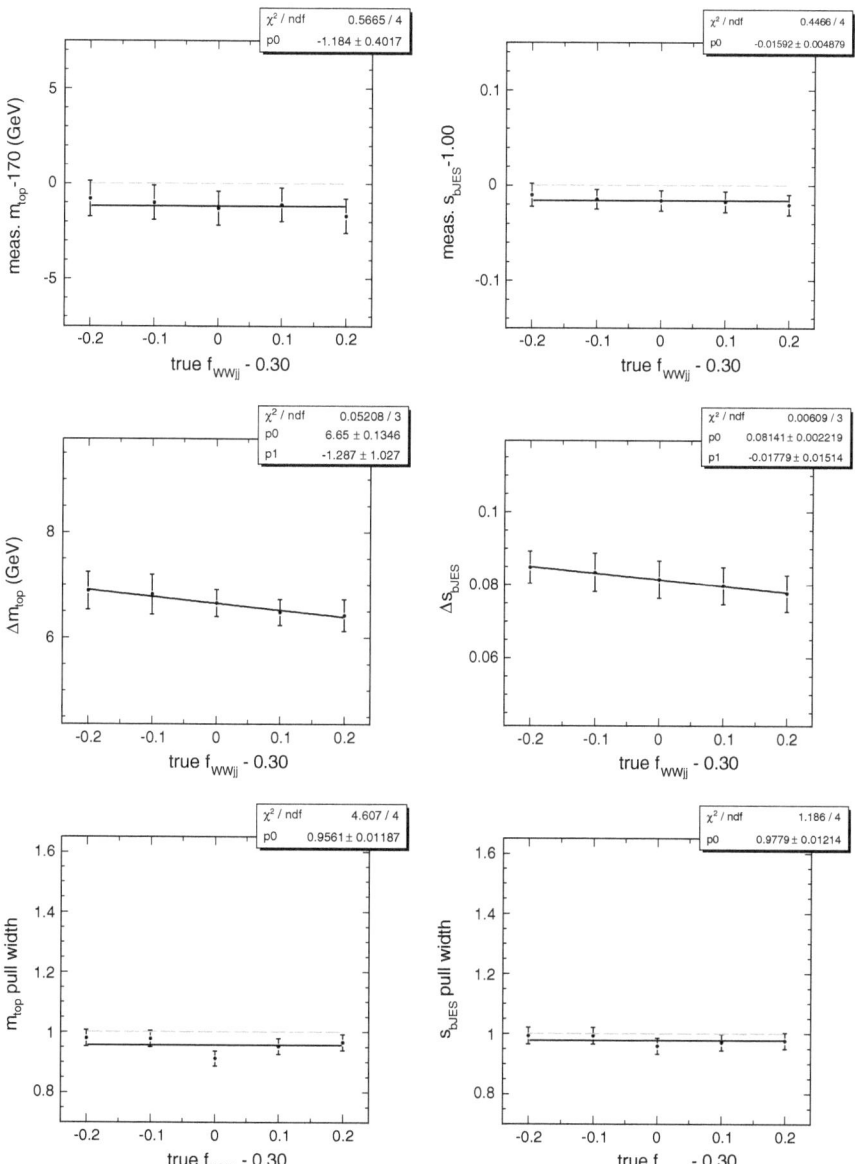

Fig. 7.10 *Left*: measurement of the top quark mass, its uncertainty, and its pull width as a function of the *WWjj* fraction in background events; *right*: measurement of the *b* jet energy scale, its uncertainty, and its pull width as a function of the *WWjj* fraction in background events. The signal fraction is fixed to 70%

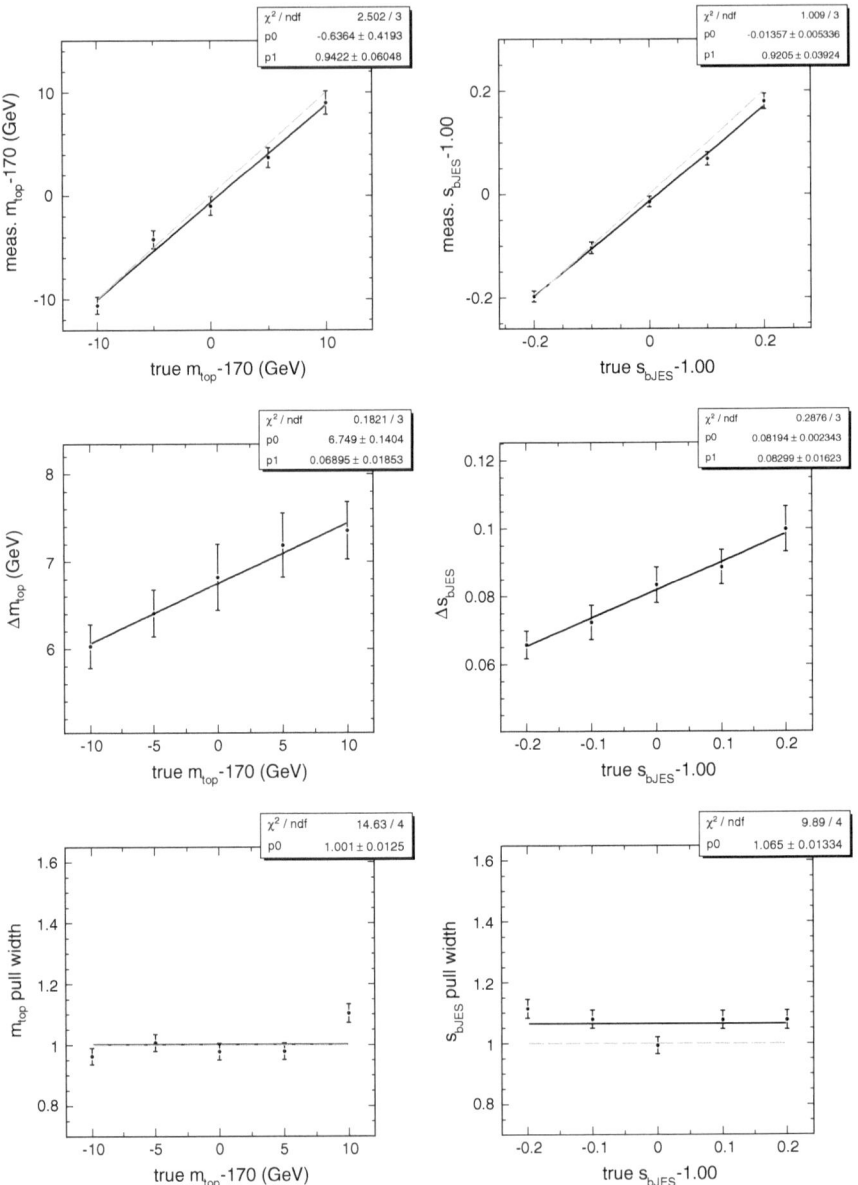

Fig. 7.11 *Left*: measurement of the top quark mass, its uncertainty, and its pull width as a function of the top quark mass; *right*: measurement of the *b* jet energy scale, its uncertainty, and its pull width as a function of the *b* jet energy scale. On average, 70% $t\bar{t}$, 24% $(Z \rightarrow \tau\tau)jj$, and 6% $WWjj$ events are used

7.3.5 Studies Including $(Z \rightarrow \tau\tau)\,jj$ and WWjj Events

Additional contamination of the $t\bar{t}$ data sample comes from $WWjj$ events. The expected fraction in the electron + muon channel compared to $(Z \rightarrow \tau\tau)\,jj$ events is about a third.

On the left side of Fig. 7.10, the measured top quark mass, its uncertainty, and its pull width as a function of the $WWjj$ fraction relative to the total number of background events in the ensembles are depicted. On the right side, the same is done for the b jet energy scale. The signal fraction is fixed to 70%, and the fraction of $WWjj$ events relative to the total number of background events is varied from 10% to 50% in steps of 10%. On average, the measured top quark mass is shifted by -1.2 GeV, the b jet energy scale by -0.02. This can be explained since no proper $WWjj$ likelihood is included.

Figure 7.11 shows the calibration curve of the top quark mass and the b jet energy scale. On average, each of the 1000 pseudo-experiments is composed of 35 signal, 10.5 $(Z \rightarrow \tau\tau)\,jj$, and 4.5 $WWjj$ events. A small dependency of m_{top} and s_{bJES} is observed, as the slope degrades to about $94 \pm 6\%$ and $92 \pm 4\%$, respectively. Within statistical uncertainties, the same effect is observed in the parton-level studies for the measurement of the top quark mass where the slope is $89 \pm 4\%$, see Sect. 6.2.5. It can be explained by the low invariant mass from leptons and jets in selected $WWjj$ events compared to invariant top quark masses of 170 or 180 GeV. The pull width stays the same within uncertainties for all calibration points.

References

1. Abazov VM et al (2006) (DØ collaboration), Measurement of the top quark mass in the lepton + jets final state with the Matrix Element method. Phys Rev D 74:092005
2. Haefner P (2008) Measurement of the top quark mass with the Matrix Element method in the semileptonic decay channel at DØ. Dissertationsschrift, Ludwig-Maximilians-Universität München
3. Maltoni F, Stelzer T (2003) MadEvent: automatic event generation with MadGraph. JHEP 302:27

Chapter 8
Conclusion

This section summarizes the results presented in this thesis, and the implications for the Standard Model of particle physics are discussed. Finally, prospects for future measurements with the Matrix Element method are outlined.

8.1 Summary and Interpretation

Since the discovery of the top quark in 1995, its mass has been measured with ever higher precision. This thesis describes the first measurement of the top quark mass in the dilepton channel using the Matrix Element method at the DØ experiment. The Matrix Element method has been developed during Run I in the lepton + jets channel when only a small data sample was available and sophisticated techniques were needed to measure the top mass with good precision.

To make use of this method in the dilepton channel, significant modifications and improvements were required. As both W bosons decay into a charged lepton and the corresponding neutrino, two out of the six final-state particles are undetected, and the reconstruction of the top quark decay is more challenging than in the lepton + jets channel. The choice of integration variables is crucial for a precise measurement and a bearable computing time. As the data sample is small in the dilepton channel, it is not desirable to remove events with additional jets as it is done in the lepton + jets analysis. On the other hand, events with more than two jets lead to a bias of about 2 GeV in the calibration curve when the top pair transverse momentum is not accounted for. Hence, the method is extended with two additional integration variables to allow for a modeling of additional jets in the event. With this improved technique, the statistical uncertainty could be reduced by about 25%. As the dominant background arises from $(Z \rightarrow \tau\tau)\,jj$ events which include the decay of a τ lepton into an electron or muon and two additional neutrinos, the design of a new process likelihood was needed. The background likelihood is calculated with the

A. Grohsjean, *Measurement of the Top Quark Mass in the Dilepton Final State Using the Matrix Element Method*, Springer Theses,
DOI: 10.1007/978-3-642-14070-9_8, © Springer-Verlag Berlin Heidelberg 2010

VECBOS matrix element, and the decay of the τ lepton is modeled using an additional transfer function that describes the energy of the charged lepton after the τ decay.

As the Matrix Element method was applied for the first time in the dilepton channel at the DØ experiment, it was validated on parton-level events demonstrating the excellent performance of the method. For the measurement, calibration curves were derived from fully simulated Monte Carlo events including detector simulation. The analysis made use of the full Run II data set recorded between April 2002 and May 2008 corresponding to an integrated luminosity of 2.8 fb^{-1}. To reduce the contamination from background events and to achieve a good agreement between simulated and measured data, a kinematic selection was applied, leaving 107 $t\bar{t}$ candidate events with one electron and one muon in the final state.

Applying the Matrix Element method to this data set, the top quark mass is measured to be

$$m_{top}^{Run~IIa} = 170.6 \pm 6.1(\text{stat.})_{-1.5}^{+2.1}(\text{syst.})\text{GeV} \tag{8.1}$$

$$m_{top}^{Run~IIb} = 174.1 \pm 4.4(\text{stat.})_{-1.8}^{+2.5}(\text{syst.})\text{GeV} \tag{8.2}$$

$$m_{top}^{comb} = 172.9 \pm 3.6(\text{stat.}) \pm 2.3(\text{syst.})\text{GeV}. \tag{8.3}$$

With a statistical uncertainty of ±3.6 GeV, this measurement has the smallest statistical uncertainty among all top mass measurements performed in the dilepton channel at the DØ experiment. The statistical uncertainty of the combined

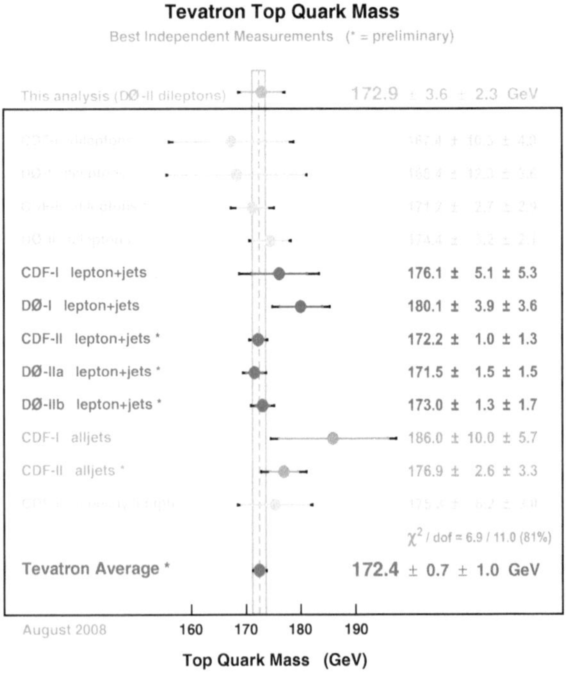

Fig. 8.1 Combination of the Run I and Run II measurements of the top quark mass at the CDF and DØ experiments in the lepton + jets and dilepton decay channel, as well as the current world average

Tevatron Top Quark Mass
Best Independent Measurements (* = preliminary)

This analysis (DØ -II dileptons)		172.9 ± 3.6 ± 2.3 GeV
CDF-I lepton+jets		176.1 ± 5.1 ± 5.3
DØ-I lepton+jets		180.1 ± 3.9 ± 3.6
CDF-II lepton+jets *		172.2 ± 1.0 ± 1.3
DØ-IIa lepton+jets *		171.5 ± 1.5 ± 1.5
DØ-IIb lepton+jets *		173.0 ± 1.3 ± 1.7
CDF-I alljets		186.0 ± 10.0 ± 5.7
CDF-II alljets *		176.9 ± 2.6 ± 3.3
		χ^2 / dof = 6.9 / 11.0 (81%)
Tevatron Average *		172.4 ± 0.7 ± 1.0 GeV

August 2008   160   170   180   190

Top Quark Mass (GeV)

measurement of the Matrix Weighting and the Neutrino Weighting methods added up to ±5.4 GeV [1]. These measurements were based on 1 fb^{-1} and included all possible dilepton final states requiring at least one lepton and one track. To allow for a direct comparison, the expected statistical uncertainties from Run IIa measurements in the electron + muon channel can be compared. Using the Neutrino Weighting technique, the average statistical uncertainty from Monte Carlo based ensemble tests was ±7.9 GeV [2], while it is only ±5.3 GeV in the Matrix Element method. Thus, a significant improvement could be achieved. One reason for the increased statistical sensitivity of this method are the transfer functions that provide an excellent mapping of the measured particles to the partonic states. In addition, the calculation of an event-per-event likelihood allows well-measured events to contribute more information to the overall likelihood than poorly measured events. Moreover, the computation of the matrix element involves the full kinematics of the event, where other methods rely only on single quantities.

The result presented has been combined with other measurements from the CDF and DØ experiments [3]. The single results from the individual analyses as well as the new world average are shown in Fig. 8.1.

Figure 8.1 shows that the top quark masses measured in the various final states agree well with each other. Moreover, the $t\bar{t}$ production cross section calculated from the combined mass value is consistent with the theoretical cross sections, compare Fig. 4.10. Finally, the relative cross sections for the different decay channels are in good agreement with each other [4]. Thus, by now there is no effect from physics beyond the Standard Model observed, and the top quark mass can be used to infer the mass of the yet-unobserved Standard Model Higgs boson. The result is shown in Fig. 8.2. The blue line depicts the relation between the two masses from Standard Model fits, where the direct top mass measurements are not included. The projection of the blue line onto the top mass axis corresponds to the predicted top mass which is in good agreement with the green band from the direct measurements. The yellow area shows the excluded mass region for the SM Higgs

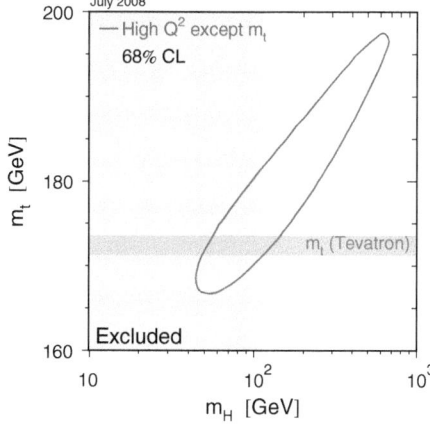

Fig. 8.2 Comparison between the direct measurement of the top quark mass (*horizontal band*) and indirect constraints on the top quark and Higgs boson from the Standard Model excluding the direct top mass measurement (*solid line*). The *left area* shows the excluded mass region for the Standard Model Higgs boson from the LEP experiments

boson from the LEP experiments. Taken this together, a Higgs boson of high mass within the Standard Model is excluded.

In a future update of this measurement, larger data sets will be available, and the statistical uncertainties will be reduced significantly. Thus, systematic uncertainties will become more and more important. With about 2 GeV, the largest systematic uncertainty arises from the uncertainty on the jet energy scale. To address this uncertainty, the Matrix Element method can be extended to simultaneously measure the top quark mass and the b jet energy scale. The adjustments needed and a large variety of parton-level tests have been discussed in this thesis. Even though the constraints on the jet energy scale of b jets are much weaker than the one for light-quark jets, and the correlation between the top quark mass and the b jet energy scale is about 0.55, a simultaneous measurement will help to reduce the total uncertainty in the future. Figure 8.3 shows the quadratic sum of the statistical uncertainty and the jet energy scale uncertainty as a function of the integrated luminosity in the dilepton channel when measuring the top quark mass only or when performing a simultaneous measurement of m_{top} and s_{bJES}, respectively. The top mass uncertainty due to the jet energy scale is assumed to be 2 GeV, see Table 6.8 in Chap. 6, the residual uncertainty due to s_{bJES}, to be 150 MeV [5]. The statistical uncertainties are estimated from the parton-level tests, see Fig. 6.13 and Fig. 7.11. From an integrated luminosity of 6 fb^{-1}, the simultaneous measurement helps to reduce the total uncertainty even when no external information on s_{bJES} is included. With the final luminosity of the Tevatron accelerator of about 10 fb^{-1}, the total uncertainty could be reduced by 20% to a value below the current systematic limit.

8.2 Outlook

In this analysis, several improvements have been introduced that are essential for a future application of the Matrix Element method at the LHC. As the phase space

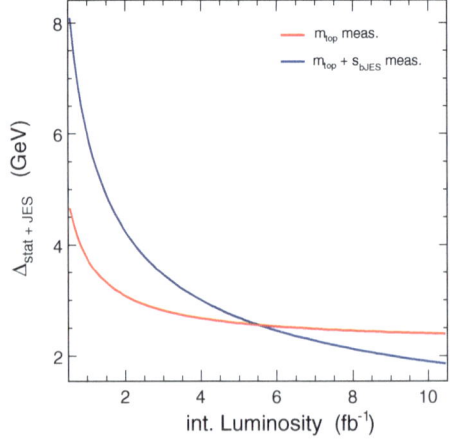

Fig. 8.3 Combined uncertainty of the statistical uncertainty and the jet energy scale uncertainty as a function of the integrated luminosity in the dilepton channel when measuring the top quark mass only (*red curve*) or when performing a simultaneous measurement of m_{top} and s_{bJES} (*blue curve*), respectively

for the production of additional jets is significantly larger at the LHC than at the Tevatron, their effect on the top quark mass measurement will be even more pronounced. The integration over the top pair transverse momentum provides a possibility to address this effect. In addition, with the two-dimensional extension of the method, a good handling of the jet energy scale uncertainty is made available.

However, as the production cross section for top quark pairs is about two orders of magnitude larger at the LHC, much larger data samples will be available within a relatively short timescale. Thus, the computing time required can easily exceed a reasonable limit. Even now, with parts of this analysis run on the computing grid, computing time is an important drawback of the method. Thus, the future of the Matrix Element method lies in measurements based on small data samples. Assuming new particles will be discovered at the LHC, the Matrix Element method constitutes a promising technique for measurements of their masses.

On the other hand, the Matrix Element method is not restricted to the measurement of a mass or energy scale. The method represents a general approach to extract information about any set of parameters in the event likelihood. For example, the Matrix Element method can be used for a measurement of spin correlations in top pair production. Even though top quarks are produced unpolarized, their spins are expected to have a strong event-by-event correlation [6] and to point along the same axis in the $t\bar{t}$ rest frame [7]. In an optimized spin quantization base, contributions from opposite spin projections are suppressed at tree level [8], and only like-sign spins occur. This measurement would be especially interesting for several reasons. As the lifetime of the top quark is proportional to the Kobayashi–Maskawa matrix element $|V_{tb}|^2$, the observation of spin correlation would yield information about the lower limit on $|V_{tb}|$, without assuming that there are three generations of quark families [9]. Moreover, it allows to probe for the production and decay dynamics of the top quark, and it is an excellent test of the Standard Model.

Moreover, the Matrix Element method can be used to measure the helicity of W bosons in top decays [10]. In the limit of a massless b quark, the standard V-A coupling at the top decay vertex requires that the produced b quark is left-handed, restricting the helicity of the W^+ boson to values of 0 and -1. The observation of a different helicity component would clearly indicate the presence of physics beyond the Standard Model.

References

1. Abazov VM et al (2007) Combination of preliminary dilepton mass measurements by the DØ experiment. DØ note 5460
2. Kehoe R et al (2008) Measurement of the top quark mass in the dilepton channel via Neutrino Weighting. DØ note 5754
3. The Tevatron Electroweak Working Group for the CDF and DØ collaborations (2008) Combination of CDF and DØ results on the mass of the top quark. arXiv:hep-ex/0703034

4. Abazov VM et al (2008) (DØ collaboration), A search for charged Higgs bosons in $t\bar{t}$ events. DØ note 5715

5. Haefner P (2008) Measurement of the top quark mass with the Matrix Element method in the semileptonic decay channel at DØ. Dissertationsschrift, Ludwig-Maximilians-Universität München

6. Barger V et al (1998) Spin correlation effects in the hadroproduction and decay of very heavy top quark pairs. Int J Mod Phys A 4:617

7. Mahlon G, Parke S (1996) Spin correlations in top quark pair production. Phys Rev D 53:4886

8. Mahlon G, Parke S (1997) Maximizing spin correlations in top quark pair production at the Tevatron. Phys Rev Lett 80:2063

9. Stelzer T, Willenbrock S (1996) Spin correlations in top quark pair production at hadron colliders. Phys Lett B 374:169

10. Abazov VM et al (2005) (DØ collaboration), Helicity of the W boson in lepton + jets $t\bar{t}$ events. Phys Lett B 617:1

Appendices

A: Solving for the Event Kinematics

As stated in Sect. 5.5, the 4-momenta of the $t\bar{t}$ decay products have to be calculated from the values of the integration variables $|\vec{p}_{b_1}|$, $|\vec{p}_{b_2}|$, $p_x^{\nu_1} - p_x^{\nu_2}$, $p_y^{\nu_1} - p_y^{\nu_2}$, $p_x^{t\bar{t}}$, $p_y^{t\bar{t}}$, $m_{\mathrm{top}_1}^2$, $m_{\mathrm{top}_2}^2$ and $(q/p_T)_{\mu_1}$ and $(q/p_T)_{\mu_2}$ (for top decays involving muons), the measured jet and lepton angles, and the electron energies (in the electron case). The calculation is derived in this section.

In the following, p_a is the 4-momentum of particle a, and similarly m_a, E_a, and $|\vec{p}_a|$ are its mass, energy, and momentum magnitude in the laboratory frame, respectively. The symbol p_{ab} is defined as the sum of the 4-momenta of particles a and b; corresponding notations are used for their energy and the magnitude of the sum of their momenta.

The neutrino transverse momentum components can be obtained as follows:

$$p_x^{\nu_1} = \frac{1}{2}\left(\not{E}_x + \Delta_{p_x^\nu}\right) \text{ and} \tag{A.1}$$

$$p_x^{\nu_2} = \frac{1}{2}\left(\not{E}_x - \Delta_{p_x^\nu}\right) \tag{A.2}$$

where

$$\not{E}_x = -\left(-p_x^{t\bar{t}} + p_x^{b_1} + p_x^{\ell_1} + p_x^{b_2} + p_x^{\ell_2}\right) \tag{A.3}$$

$$\Delta_{p_x^\nu} = p_x^{\nu_1} - p_x^{\nu_2}. \tag{A.4}$$

Note that this is not the same as the x component of the measured missing transverse energy which is not used in the measurement.

The longitudinal neutrino momenta are calculated as follows using the top masses (the calculation is done in the same way for both neutrinos):

$$m_{\text{top}}^2 = (p_\ell + p_b + p_v)^2$$

$$= m_{b\ell}^2 + 2p_{b\ell}p_v$$

$$= m_{b\ell}^2 + 2E_{b\ell}E_v - 2\vec{p}_{b\ell} \cdot \vec{p}_v \tag{A.5}$$

$$\Leftrightarrow \frac{1}{2}\left(m_{\text{top}}^2 - m_{b\ell}^2\right) = E_{b\ell}E_v - \vec{p}_{b\ell} \cdot \vec{p}_v$$

$$\Leftrightarrow \frac{1}{2}\left(m_{\text{top}}^2 - m_{b\ell}^2\right) + p_x^{b\ell}p_x^v + p_y^{b\ell}p_y^v + p_z^{b\ell}p_z^v = E_{b\ell}\sqrt{p_x^{v2} + p_y^{v2} + p_z^{v2}}. \tag{A.6}$$

Defining

$$A := \left(\frac{1}{2}\left(m_{\text{top}}^2 - m_{b\ell}{}^2\right) + p_x^{b\ell}p_x^v + p_y^{b\ell}p_y^v\right)/E_{b\ell} \tag{A.7}$$

and squaring the equation, it follows that

$$\Rightarrow \left(A + \frac{p_z^{b\ell}}{E_{b\ell}}p_z^v\right)^2 = p_x^{v2} + p_y^{v2} + p_z^{v2} \tag{A.8}$$

$$\Leftrightarrow A^2 - p_x^{v2} - p_y^{v2} + 2A\frac{p_z^{b\ell}}{E_{b\ell}}p_z^v = \left(1 - \left(\frac{p_z^{b\ell}}{E_{b\ell}}\right)^2\right)p_z^{v2},$$

and with

$$B := \frac{A^2 - p_x^{v2} - p_y^{v2}}{1 - \left(\frac{p_z^{b\ell}}{E_{b\ell}}\right)^2} \text{ and} \tag{A.9}$$

$$C := \frac{A\frac{p_z^{b\ell}}{E_{b\ell}}}{1 - \left(\frac{p_z^{b\ell}}{E_{b\ell}}\right)^2}, \tag{A.10}$$

one gets

$$B + 2Cp_z^v = p_z^{v2}$$

$$\Leftrightarrow B + C^2 = \left(p_z^v - C\right)^2 \tag{A.11}$$

$$\Leftrightarrow p_z^v = C \pm \sqrt{B + C^2}.$$

Because of the transition from equation (A.6) to the squared equation (A.8) it must be checked whether the solutions in equation (A.11) actually solve the initial equation (A.5), otherwise the point in integration space does not contribute to the integral that yields the signal probability.

The energies of the colliding partons are given by

$$E_\pm = \frac{1}{2}\left(E_{b\ell^+\bar\nu b\ell^-\bar\nu} \pm p_z^{b\ell^+\bar\nu b\ell^-\bar\nu}\right) . \qquad (A.12)$$

It has to be checked that both parton energies are smaller than the beam energy.

B: The Jacobian Determinant for the Signal Integration

In the following, the Jacobian determinant for the transition from an integral over parton momenta to the integration variables $|\vec{p}_{b_1}|$, $|\vec{p}_{b_2}|$, $p_x^{\nu_1}-p_x^{\nu_2}$, $p_y^{\nu_1}-p_y^{\nu_2}$, $m_{\text{top}_1}^2$, and $m_{\text{top}_2}^2$ is derived. As a starting point, it is assumed that the neutrino momenta are given in Cartesian coordinates, while the momenta of the two b quarks are given in spherical coordinates (the transformation from Cartesian to spherical coordinates for the b quark momenta adds additional Jacobian factors, see below). It is further assumed that the integration over the δ distributions describing perfect measurements, like the b quark and lepton directions and electron energies, has already been performed such that the corresponding variables need no longer to be considered here. First, the case of zero $t\bar{t}$ transverse momentum is considered, and this condition is used to relate the transverse momentum components of the two neutrinos and eliminate $p_x^{\nu_2}$ and $p_y^{\nu_2}$. The discussion below concentrates on the case of the ee final state. This means that the additional complication of the muon momentum integration is discussed separately further below.

The determinant for the above-mentioned transformation is

$$\det(J) = \det \begin{pmatrix}
\dfrac{\partial|\vec{p}_{b_1}|}{\partial|\vec{p}_{b_1}|} & \dfrac{\partial|\vec{p}_{b_2}|}{\partial|\vec{p}_{b_1}|} & \dfrac{\partial(p_x^{\nu_1}-p_x^{\nu_2})}{\partial|\vec{p}_{b_1}|} & \dfrac{\partial(p_y^{\nu_1}-p_y^{\nu_2})}{\partial|\vec{p}_{b_1}|} & \dfrac{\partial m_{\text{top}_1}^2}{\partial|\vec{p}_{b_1}|} & \dfrac{\partial m_{\text{top}_2}^2}{\partial|\vec{p}_{b_1}|} \\[2.2ex]
\dfrac{\partial|\vec{p}_{b_1}|}{\partial|\vec{p}_{b_2}|} & \dfrac{\partial|\vec{p}_{b_2}|}{\partial|\vec{p}_{b_2}|} & \dfrac{\partial(p_x^{\nu_1}-p_x^{\nu_2})}{\partial|\vec{p}_{b_2}|} & \dfrac{\partial(p_y^{\nu_1}-p_y^{\nu_2})}{\partial|\vec{p}_{b_2}|} & \dfrac{\partial m_{\text{top}_1}^2}{\partial|\vec{p}_{b_2}|} & \dfrac{\partial m_{\text{top}_2}^2}{\partial|\vec{p}_{b_2}|} \\[2.2ex]
\dfrac{\partial|\vec{p}_{b_1}|}{\partial p_x^{\nu_1}} & \dfrac{\partial|\vec{p}_{b_2}|}{\partial p_x^{\nu_1}} & \dfrac{\partial(p_x^{\nu_1}-p_x^{\nu_2})}{\partial p_x^{\nu_1}} & \dfrac{\partial(p_y^{\nu_1}-p_y^{\nu_2})}{\partial p_x^{\nu_1}} & \dfrac{\partial m_{\text{top}_1}^2}{\partial p_x^{\nu_1}} & \dfrac{\partial m_{\text{top}_2}^2}{\partial p_x^{\nu_1}} \\[2.2ex]
\dfrac{\partial|\vec{p}_{b_1}|}{\partial p_y^{\nu_1}} & \dfrac{\partial|\vec{p}_{b_2}|}{\partial p_y^{\nu_1}} & \dfrac{\partial(p_x^{\nu_1}-p_x^{\nu_2})}{\partial p_y^{\nu_1}} & \dfrac{\partial(p_y^{\nu_1}-p_y^{\nu_2})}{\partial p_y^{\nu_1}} & \dfrac{\partial m_{\text{top}_1}^2}{\partial p_y^{\nu_1}} & \dfrac{\partial m_{\text{top}_2}^2}{\partial p_y^{\nu_1}} \\[2.2ex]
\dfrac{\partial|\vec{p}_{b_1}|}{\partial p_z^{\nu_1}} & \dfrac{\partial|\vec{p}_{b_2}|}{\partial p_z^{\nu_1}} & \dfrac{\partial(p_x^{\nu_1}-p_x^{\nu_2})}{\partial p_z^{\nu_1}} & \dfrac{\partial(p_y^{\nu_1}-p_y^{\nu_2})}{\partial p_z^{\nu_1}} & \dfrac{\partial m_{\text{top}_1}^2}{\partial p_z^{\nu_1}} & \dfrac{\partial m_{\text{top}_2}^2}{\partial p_z^{\nu_1}} \\[2.2ex]
\dfrac{\partial|\vec{p}_{b_1}|}{\partial p_z^{\nu_2}} & \dfrac{\partial|\vec{p}_{b_2}|}{\partial p_z^{\nu_2}} & \dfrac{\partial(p_x^{\nu_1}-p_x^{\nu_2})}{\partial p_z^{\nu_2}} & \dfrac{\partial(p_y^{\nu_1}-p_y^{\nu_2})}{\partial p_z^{\nu_2}} & \dfrac{\partial m_{\text{top}_1}^2}{\partial p_z^{\nu_2}} & \dfrac{\partial m_{\text{top}_2}^2}{\partial p_z^{\nu_2}}
\end{pmatrix} \qquad (B.1)$$

$$= \det \begin{pmatrix}
1 & 0 & \dfrac{\partial\left(p_x^{v1}-p_x^{v2}\right)}{\partial|\vec{p}_{b_1}|} & \dfrac{\partial\left(p_y^{v1}-p_y^{v2}\right)}{\partial|\vec{p}_{b_1}|} & \dfrac{\partial m^2_{\text{top}_1}}{\partial|\vec{p}_{b_1}|} & \dfrac{\partial m^2_{\text{top}_2}}{\partial|\vec{p}_{b_1}|} \\[2mm]
0 & 1 & \dfrac{\partial\left(p_x^{v1}-p_x^{v2}\right)}{\partial|\vec{p}_{b_2}|} & \dfrac{\partial\left(p_y^{v1}-p_y^{v2}\right)}{\partial|\vec{p}_{b_2}|} & \dfrac{\partial m^2_{\text{top}_1}}{\partial|\vec{p}_{b_2}|} & \dfrac{\partial m^2_{\text{top}_2}}{\partial|\vec{p}_{b_2}|} \\[2mm]
0 & 0 & 2 & 0 & \dfrac{\partial m^2_{\text{top}_1}}{\partial p_x^{v1}} & \dfrac{\partial m^2_{\text{top}_2}}{\partial p_x^{v1}} \\[2mm]
0 & 0 & 0 & 2 & \dfrac{\partial m^2_{\text{top}_1}}{\partial p_y^{v1}} & \dfrac{\partial m^2_{\text{top}_2}}{\partial p_y^{v1}} \\[2mm]
0 & 0 & 0 & 0 & \dfrac{\partial m^2_{\text{top}_1}}{\partial p_z^{v1}} & 0 \\[2mm]
0 & 0 & 0 & 0 & 0 & \dfrac{\partial m^2_{\text{top}_2}}{\partial p_z^{v2}}
\end{pmatrix}$$

$$= 4\,\frac{\partial m^2_{\text{top}_1}}{\partial p_z^{v1}}\,\frac{\partial m^2_{\text{top}_2}}{\partial p_z^{v2}}$$

The derivatives relevant for the determinant can be derived as follows:

$$m^2_{\text{top}} = (p_b + p_\ell + p_v)^2 \tag{B.2}$$

$$= m_{b\ell}^2 + 2E_{b\ell}E_v - 2\vec{p}_{b\ell}\cdot\vec{p}_v \tag{B.3}$$

$$\Leftrightarrow \frac{\partial m^2_{\text{top}}}{\partial p_z^v} = 2E_{b\ell}\frac{\partial E_v}{\partial p_z^v} - 2p_z^{b\ell} \tag{B.4}$$

$$= 2\left(E_{b\ell}u_z^v - p_z^{b\ell}\right) \tag{B.5}$$

where in the last transition,

$$\frac{\partial\sqrt{p_x^{v2}+p_y^{v2}+p_z^{v2}}}{\partial p_z^v} = \frac{p_z^v}{E_v} = u_z^v \tag{B.6}$$

has been used.

As a second step, the more general case of non-zero $t\bar{t}$ transverse momentum is considered. Here, two additional integration variables, $p_x^{t\bar{t}}$ and $p_y^{t\bar{t}}$, have to be introduced.

The Jacobian determinant becomes

$$
\det(J') = \det
\begin{pmatrix}
\dfrac{\partial|\vec{p}_{b_1}|}{\partial|\vec{p}_{b_1}|} & \dfrac{\partial|\vec{p}_{b_2}|}{\partial|\vec{p}_{b_1}|} & \dfrac{\partial(p_x^{\nu_1}-p_x^{\nu_2})}{\partial|\vec{p}_{b_1}|} & \dfrac{\partial(p_y^{\nu_1}-p_y^{\nu_2})}{\partial|\vec{p}_{b_1}|} & \dfrac{\partial m_{\mathrm{top}_1}^2}{\partial|\vec{p}_{b_1}|} & \dfrac{\partial m_{\mathrm{top}_2}^2}{\partial|\vec{p}_{b_1}|} & \dfrac{\partial p_x^{t\bar{t}}}{\partial|\vec{p}_{b_1}|} & \dfrac{\partial p_y^{t\bar{t}}}{\partial|\vec{p}_{b_1}|} \\[2.2ex]
\dfrac{\partial|\vec{p}_{b_1}|}{\partial|\vec{p}_{b_2}|} & \dfrac{\partial|\vec{p}_{b_2}|}{\partial|\vec{p}_{b_2}|} & \dfrac{\partial(p_x^{\nu_1}-p_x^{\nu_2})}{\partial|\vec{p}_{b_2}|} & \dfrac{\partial(p_y^{\nu_1}-p_y^{\nu_2})}{\partial|\vec{p}_{b_2}|} & \dfrac{\partial m_{\mathrm{top}_1}^2}{\partial|\vec{p}_{b_2}|} & \dfrac{\partial m_{\mathrm{top}_2}^2}{\partial|\vec{p}_{b_2}|} & \dfrac{\partial p_x^{t\bar{t}}}{\partial|\vec{p}_{b_2}|} & \dfrac{\partial p_y^{t\bar{t}}}{\partial|\vec{p}_{b_2}|} \\[2.2ex]
\dfrac{\partial|\vec{p}_{b_1}|}{\partial p_x^{\nu_1}} & \dfrac{\partial|\vec{p}_{b_2}|}{\partial p_x^{\nu_1}} & \dfrac{\partial(p_x^{\nu_1}-p_x^{\nu_2})}{\partial p_x^{\nu_1}} & \dfrac{\partial(p_y^{\nu_1}-p_y^{\nu_2})}{\partial p_x^{\nu_1}} & \dfrac{\partial m_{\mathrm{top}_1}^2}{\partial p_x^{\nu_1}} & \dfrac{\partial m_{\mathrm{top}_2}^2}{\partial p_x^{\nu_1}} & \dfrac{\partial p_x^{t\bar{t}}}{\partial p_x^{\nu_1}} & \dfrac{\partial p_y^{t\bar{t}}}{\partial p_x^{\nu_1}} \\[2.2ex]
\dfrac{\partial|\vec{p}_{b_1}|}{\partial p_y^{\nu_1}} & \dfrac{\partial|\vec{p}_{b_2}|}{\partial p_y^{\nu_1}} & \dfrac{\partial(p_x^{\nu_1}-p_x^{\nu_2})}{\partial p_y^{\nu_1}} & \dfrac{\partial(p_y^{\nu_1}-p_y^{\nu_2})}{\partial p_y^{\nu_1}} & \dfrac{\partial m_{\mathrm{top}_1}^2}{\partial p_y^{\nu_1}} & \dfrac{\partial m_{\mathrm{top}_2}^2}{\partial p_y^{\nu_1}} & \dfrac{\partial p_x^{t\bar{t}}}{\partial p_y^{\nu_1}} & \dfrac{\partial p_y^{t\bar{t}}}{\partial p_y^{\nu_1}} \\[2.2ex]
\dfrac{\partial|\vec{p}_{b_1}|}{\partial p_z^{\nu_1}} & \dfrac{\partial|\vec{p}_{b_2}|}{\partial p_z^{\nu_1}} & \dfrac{\partial(p_x^{\nu_1}-p_x^{\nu_2})}{\partial p_z^{\nu_1}} & \dfrac{\partial(p_y^{\nu_1}-p_y^{\nu_2})}{\partial p_z^{\nu_1}} & \dfrac{\partial m_{\mathrm{top}_1}^2}{\partial p_z^{\nu_1}} & \dfrac{\partial m_{\mathrm{top}_2}^2}{\partial p_z^{\nu_1}} & \dfrac{\partial p_x^{t\bar{t}}}{\partial p_z^{\nu_1}} & \dfrac{\partial p_y^{t\bar{t}}}{\partial p_z^{\nu_1}} \\[2.2ex]
\dfrac{\partial|\vec{p}_{b_1}|}{\partial p_z^{\nu_2}} & \dfrac{\partial|\vec{p}_{b_2}|}{\partial p_z^{\nu_2}} & \dfrac{\partial(p_x^{\nu_1}-p_x^{\nu_2})}{\partial p_z^{\nu_2}} & \dfrac{\partial(p_y^{\nu_1}-p_y^{\nu_2})}{\partial p_z^{\nu_2}} & \dfrac{\partial m_{\mathrm{top}_1}^2}{\partial p_z^{\nu_2}} & \dfrac{\partial m_{\mathrm{top}_2}^2}{\partial p_z^{\nu_2}} & \dfrac{\partial p_x^{t\bar{t}}}{\partial p_z^{\nu_2}} & \dfrac{\partial p_y^{t\bar{t}}}{\partial p_z^{\nu_2}} \\[2.2ex]
\dfrac{\partial|\vec{p}_{b_1}|}{\partial p_x^{\nu_2}} & \dfrac{\partial|\vec{p}_{b_2}|}{\partial p_x^{\nu_2}} & \dfrac{\partial(p_x^{\nu_1}-p_x^{\nu_2})}{\partial p_x^{\nu_2}} & \dfrac{\partial(p_y^{\nu_1}-p_y^{\nu_2})}{\partial p_x^{\nu_2}} & \dfrac{\partial m_{\mathrm{top}_1}^2}{\partial p_x^{\nu_2}} & \dfrac{\partial m_{\mathrm{top}_2}^2}{\partial p_x^{\nu_2}} & \dfrac{\partial p_x^{t\bar{t}}}{\partial p_x^{\nu_2}} & \dfrac{\partial p_y^{t\bar{t}}}{\partial p_x^{\nu_2}} \\[2.2ex]
\dfrac{\partial|\vec{p}_{b_1}|}{\partial p_y^{\nu_2}} & \dfrac{\partial|\vec{p}_{b_2}|}{\partial p_y^{\nu_2}} & \dfrac{\partial(p_x^{\nu_1}-p_x^{\nu_2})}{\partial p_y^{\nu_2}} & \dfrac{\partial(p_y^{\nu_1}-p_y^{\nu_2})}{\partial p_y^{\nu_2}} & \dfrac{\partial m_{\mathrm{top}_1}^2}{\partial p_y^{\nu_2}} & \dfrac{\partial m_{\mathrm{top}_2}^2}{\partial p_y^{\nu_2}} & \dfrac{\partial p_x^{t\bar{t}}}{\partial p_y^{\nu_2}} & \dfrac{\partial p_y^{t\bar{t}}}{\partial p_y^{\nu_2}}
\end{pmatrix}
$$

$$
= 4 \frac{\partial m_{\mathrm{top}_1}^2}{\partial p_z^{\nu_1}} \frac{\partial m_{\mathrm{top}_2}^2}{\partial p_z^{\nu_2}} \det
\begin{pmatrix}
\dfrac{\partial p_x^{t\bar{t}}}{\partial p_x^{\nu_2}} & \dfrac{\partial p_y^{t\bar{t}}}{\partial p_x^{\nu_2}} \\[2.2ex]
\dfrac{\partial p_x^{t\bar{t}}}{\partial p_y^{\nu_2}} & \dfrac{\partial p_y^{t\bar{t}}}{\partial p_y^{\nu_2}}
\end{pmatrix}
$$

$$
= 4 \frac{\partial m_{\mathrm{top}_1}^2}{\partial p_z^{\nu_1}} \frac{\partial m_{\mathrm{top}_2}^2}{\partial p_z^{\nu_2}} \det
\begin{pmatrix}
1 & 0 \\
0 & 1
\end{pmatrix}
$$

$$
= 4 \frac{\partial m_{\mathrm{top}_1}^2}{\partial p_z^{\nu_1}} \frac{\partial m_{\mathrm{top}_2}^2}{\partial p_z^{\nu_2}}
$$

(B.7)

and is thus equal to the determinant in the previous, simpler case.

The additional factors to be included because the transfer functions are determined as a function of jet energy and inverse muon transverse momentum are identical to those in the lepton + jets analysis [1].

It should be noted that for the numerical integration performed here, where the integral is always computed from the lower to the upper boundary irrespective of the sign of the Jacobian determinant, the absolute value of the determinant has to be included as a factor in the integration.

The total factor to be included in the integration is thus

$$\left| \frac{1}{\det(J)} \right| \tag{B.8}$$

with $\det(J)$ given in (B.1).

Reference

1. Schieferdecker P (2005) Measurement of the top quark mass at DØ Run II with the Matrix Element method in the lepton+jets final state. Dissertationsschrift, Ludwig-Maximilians-Universität München

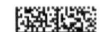